Keeping Sheep

By the same author

KEEPING GOATS
KEEPING RABBITS
KEEPING PIGS

Garden Farming Series

Keeping Sheep

Elisabeth Downing

PELHAM BOOKS

First published in Great Britain by
PELHAM BOOKS LTD
52 Bedford Square
London WC1B 3EF
1979

ISBN 0 7207 1145 2

Filmset and printed in Great Britain by
BAS Printers Limited, Over Wallop, Hampshire

Contents

'. . . ewes, when taken great care of, will be very
gentle . . .
they must be fed with nice and clean food . . .'

William Cobbett (1762–1835)
Cottage Economy

Preface

Why do we want to keep sheep? The irresistible aroma of roast leg of lamb may drive some of us to utilize the odd grassy areas in our gardens. On the other hand the yearly harvest of fluffy fleeces asking to be carded, spun, possibly dyed and then knitted or woven into practical things like clothes, carpets and curtains may inspire others to keep a couple of ewes who will, with luck, produce lambs for the deep freeze, maybe in cooperation with a neighbour. Tending the creatures will provide an extremely satisfying hobby; it will also inculcate in our children a knowledge of our responsibilities to domestic animals. The added bonus of increased land fertility and a possible improvement in the grass sward will be realized after the first year.

However, keeping sheep is not all sunshine. There are times when we don't feel like going out to check the fencing on a wet and cold November evening. The pregnant ewe will require an unobtrusive, tactful watch kept on her. There will be times when drought or constant rain will make us feel we never want to see another sheep. But I believe that the advantages far outweigh the pitfalls and difficulties, and I hope this book will help the newcomer to shepherding. If I discourage those who imagine that a few unattended sheep on the back lawn are an effortless way to unlimited roast lamb and kilograms of ready-spun wool, my labours will not have been in vain.

Elisabeth C. M. Downing

Acknowledgments

Thanks must go to Dulcie Asker, whose encouragement and expert typing of the final manuscript were invaluable. I must also thank my long-suffering husband, who spent hours deciphering and typing the initial copy, annotated the figures and once more supplied the cover photograph.

Many shepherds gave valuable time and advice for which I am most grateful, including Mrs Patricia Russell, Colonel Peter and Mrs Halliday, Mrs Mary Lambert and Mrs Jo Baxter.

I am particularly indebted to Zuleika Ainley of the Strangers Wool Shop in Norwich for writing the chapter on 'Using Your Wool' and also to Rashida McCabe, who helped her with the section on dyeing.

Finally I must thank Lesley Gowers, my patient and understanding editor.

Conversion Table

Metric and Imperial Equivalents

Imperial	Metric	Metric	Imperial
1 inch	2.54 cm	1 cm	0.39 in
1 foot	30.48 cm	1 cm	0.033 ft
1 yard	0.91 m	1 m	1.094 yds
1 mile	1.61 km	1 km	0.62 miles
1 sq yd	0.84 sq m	1 sq m	1.196 sq yds
1 cu yd	0.76 cu m	1 cu m	1.31 cu yds
1 pint	0.57 litre	1 litre	1.76 pints
1 gal	0.0056 cu m	1 cu m	219.97 gals
1 gal	4.55 litre	1 litre	0.22 gals
1 fl oz	28.4 ml	1 ml	0.035 fl oz
1 oz	28.35 g	1 g	0.035 oz
1 lb	0.45 kg	1 kg	2.20 lb
1 acre	0.405 hectare	1 hectare	2.47 acres
$x\,°F$	$\frac{5}{9}(x-32)\,°C$	$y\,°C$	$(\frac{9}{5}y+32)\,°F$

Milk
1 lb $= \frac{3}{4}$ pint
1 pint $= 1\frac{1}{3}$ lb

Metric abbreviations
cm	centimetre
m	metre
km	kilometre
ml	millilitre
g	gram
kg	kilogram

1 Making a Start

Commercially, sheep are kept for a number of reasons, including the production of lamb for slaughter, wool and the possible sale of pedigree stock. As a garden farmer you have the luxury of choosing to keep sheep for reasons besides making money.

Two or three orphan lambs acquired from a nearby commercial flock and reared on bottles can give much pleasure and satisfaction as well as keeping down excess grass. They will also provide succulent lamb chops for the deep freeze or a juicy leg of lamb as a basis for barter. But looking at it realistically, they could prove extremely expensive if time and money spent feeding and tending them were costed out in a businesslike way. With a little more grass available, a couple or so ewe lambs could be kept with a different end in mind. At maturity a visit to a neighbouring ram may eventually produce two or more lambs, to be followed by fleeces later in the year. This wool, depending on its type, could be spun and knitted or woven into a variety of articles, the coarse wools for tweeds and carpets and the finer ones for soft sweaters and shawls.

Within the last few years there has been a widely reawakened interest in the craft of wool hand-spinning, with groups of people meeting to spin together, listening to lectures on allied subjects and giving demonstrations and displays of spinning and weaving. Others prefer to spin alone by their own firesides, as does one of my daughters! Imagine the wealth of memories brought back when sitting down in front of your own home-grown fleece, clean and neatly rolled, waiting to be sorted and spun. The warm, sheepy, tallow aroma always conjures up a feeling of comfort and pleasure in me, and I forget the drag of midnight feeds for an ailing orphan lamb and the hours spent trying to find and mend broken fencing.

On the Continent and further east, sheep have been kept for thousands of years for their milk as well as fleeces and meat. In France flocks are retained for the production of milk for Roquefort cheese; the East Friesian sheep is particularly renowned for its milking capacity and is now kept by a few interested breeders in England. It can yield up to three litres or more daily for two months after its young have been weaned at about six weeks of age, and will continue to milk well into the autumn.

The professional shepherd has more than a full-time job with his flock. But the garden farmer and his family can, with a little organization, fit in keeping a few sheep with the household duties and even a full-time job, as long as there is someone at home for some of the day. Obviously they will need attention at least twice daily, and provision will have to be made for them if everyone goes away on holiday together. Often a few sheep can be taken to like-minded friends who will know how to care for them. In our case we never seem to go away all at the same time so the problem does not arise.

Over fifty separate recognized breeds of sheep in the British Isles must present a daunting array to the uninitiated. But deeper research will show that each breed serves a particular purpose for the wide varieties of terrain, weather conditions and management that are found in different parts of this country.

In order to simplify the problem the breeds can be loosely classified as follows:

Shortwools, e.g. Dorset Horn, Wiltshire, Ryeland, Devon Closewool, Oxford, Hampshire, Southdown, Suffolk, Dorset Down. These include the various Down breeds which were originally bred for grazing on the Downlands of southern England. They

Fig. 1 Suffolk sheep, a shortwool type.

Fig. 2 Lincoln Longwool.

Fig. 3 Scots Blackface, a mountain-type sheep.

Fig. 4 Clun Forest, an intermediate type.

produce early, meaty carcases in the more fertile areas.

Longwools, e.g. Lincoln Longwool, Leicester, Blue-faced Leicester, Border Leicester, Wensleydale, Cotswold. Here the wool is slightly coarser and the fleece is both longer and heavier. The carcase also is larger.

Mountain, e.g. Scots Blackface, Welsh Mountain, Swaledale, Lonk, Cheviot, Rough Fell. These animals are small and hardy, with possibly an even coarser quality fleece suitable for making into tweeds and blankets. The carcases yield an extremely sweet mutton.

Intermediate, e.g. Clun Forest, Llanwenog, Kerry Hill, Beulah Speckled Face. These breeds originated from the mountain sheep, and are often found on the lower hills and poor, less fertile areas. They will produce early lambs to fatten on the lowlands when crossed with a shortwool ram.

The National Sheep Association produces a book, *British Sheep*, where photographs and a brief description of most of the breeds and crosses occurring can be found, together with the addresses of the breed societies and particulars of many of the regular breed sales. The British Wool Marketing Board also produce an attractive informative booklet illustrated with excellent photographs of the main breeds of sheep.

The Rare Breed Survival Trust was launched in 1973 to foster interest in the less common breeds of livestock (some of which were, and still are, facing extinction). Its work has drawn attention to the plight of some of the more primitive types of sheep, as well as various more advanced breeds whose numbers were falling dangerously low. Breed names to look out for include Manx Loghtan, St Kilda, North Ronaldsay, Soay, Whitefaced Woodland and Portland. A combined flock book has been formed offering facilities for registration for animals of these breeds. The Rare Breed Survival Trust, besides exhibiting examples of rare breeds of animals and an interesting display of literature at many of the major agricultural shows, organizes an annual show and sale with classes for cattle, sheep and goats. Breeders' Workshops are also periodically organized throughout the country. *The Ark*, which is the monthly journal of the RBST, frequently contains articles of particular concern to those interested in sheep and various aspects of management, including the production of wool for spinning and weaving.

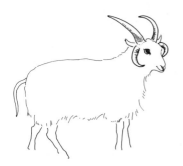

Fig. 5 Manx Loghtan, a rare breed.

Fig. 6 Jacob sheep.

Those who wish to breed a sheep which will yield a fleece providing a yarn fit for a particular purpose will have their choice of breed considerably reduced, as different breeds yield very different fibres. The thicker, coarser-fibred wools are used for rugs, tweeds and carpets, and are produced by such breeds as the Mountain Blackface, Swaledale and Rough Fell sheep, while at the other extreme we find the Shetland sheep yielding a wool noted for its fineness and softness which is widely used in the Shetland Isles for knitting shawls and warm garments. Of course individual sheep within a breed *may* grow widely differing qualities and weights of wool.

A breed which has gained much favour and whose numbers have increased enormously in the last few years is the Jacob sheep, a breed which came to England from Spain in the seventeenth century. The breed society was founded in 1969 to foster the breeding of Jacob sheep, and it sets out to improve conformation of the carcase, staple length of wool and prolificacy. This sheep produces a distinctive black and white lamb which changes to brown and white as it grows older. An attractive feature is the fact that Jacobs range from being polled (hornless) to having no less than six horns. They seem to thrive under a wide range of conditions and management, and are generally easily tamed and very amenable. The wool may be equally suitable for knitting as for mat making. Being produced in a range of colours from rich chocolate through brown to white it is widely used undyed and unbleached. The meat produced is particularly sweet and lean.

The quality of stock varies widely and the prospective buyer would be wise to study the conformation of the animal and the quality of the fleece. Generally the breed is particularly prolific – two, three and even more lambs often being produced in one litter. The ewes make excellent mothers, as well as being exceptionally milky. However, attention should to be paid to the horned animals, and those with horns tending to curl in towards the head should be avoided, as if this condition is left unattended one or more horns may eventually grow into the actual flesh; I also like to avoid those with horns pointing forwards.

Mountain breeds, though excellent mothers, tend to be highly athletic and resist attempts to restrain them within a small area. They make exceptionally hardy jumpers in both senses of the word. They prefer to graze over a wide area, and under stress will scatter. On the other hand the down and lowland breeds lend themselves to fold farming or restraint in small fields. They will usually graze close together, and are inclined to flock when frightened.

The mountain or intermediate breeds can survive under exposed conditions and on poor grazing where the shortwooled breeds would fail to flourish owing to unsuitability of fleece and insufficient high-quality keep.

Generally unless you have a strong desire to keep a particular breed and are able to supply it with its particular needs in spite of the surrounding terrain and climate (for example, *effective* fencing for mountain breeds or adequate quality food and protection for the shortwool), it is wisest to 'do in Rome as the Romans do'. Study the farms around the area and see which breeds are generally kept. Try to imagine why they have proved suitable to your locality. Make a point of visiting these farms and you will find that most shepherds will be only too delighted to talk about their own sheep.

The area and quality of grazing at your disposal must also affect your choice and number of sheep. Four head of the smaller breeds of sheep will flourish where two of the larger would probably not do very well. As a rough guideline half a hectare (just over an acre) will support four (more or less according to the breed) ewes and their lambs. Obviously poor, un-improved grass will support half the head of sheep that a well managed sward of well drained, fertile land would.

If you only intend rearing two or three orphan lambs for slaughter at about sixteen weeks of age, a much smaller area would suffice if you are prepared to change the area grazed daily. This is really a fairly simple way to keep sheep, as the animals will only be retained for a short period each year, the land will benefit from their grazing and the resultant manure, and any parasitical worms would have little time to become established. There will also be none of the problems of winter housing or protection, and the expense of hay and concentrate feeding will be kept down to a minimum. However the resulting skins will only yield car seats or rugs, as the lambs will be slaughtered before they are old enough to be shorn.

If you do not want the bother of the newborn, bottle-fed orphan a few older weaned lambs, although considerably more expensive, will still provide a good, fat carcase. They may never become as tame as bottle-fed lambs, and this will help to make the journey to the slaughterhouse less painful for the more sensitive members of the family. There can be initial problems of restraint before the weaned lambs become accustomed to their new surroundings.

Bottle-fed lambs can become a nuisance, as they will do literally anything to find their feeder. We lost our orphan lambs once: bottle time had come around again and the kitchen door was shut, so they had followed their noses and walked in through the first open door they could find and this had slammed shut behind them. We eventually found them, having hunted high and low in the surrounding fields and lanes, happily sitting on the sofa chewing the cud. A point worth noting here is that female lambs will be less bossy as they get older compared to the male. The ram lamb left uncastrated may become aggressive as the breeding season approaches, and is quite capable of knocking an adult down. It is always unwise to retain a bottle-fed ram for service. The wether (a castrated male) is usually more boisterous than the ewe lamb, whereas the latter will tend to become progressively more docile and matronly. Remember a small lamb weighing but 5 kg or so will skip about causing little damage, while the same animal at maturity weighing anything up to 100 kg or more can be quite an uncomfortable playmate.

It is just as well to know the breed or type of your lambs, so that the expected mature weight will not prove too heavy to handle. The larger breeds can create difficulties during routine hoof trimming, shearing and dipping etc. unless you are an expert shepherd. The promise of a heavy fleece from a large Lincoln Longwool may turn sour when you come to manoeuvring a 100 kg or so sheep.

Points to observe when buying stock
Newborn lambs
1   Always ensure that the lamb has received *at least*

one meal of colostrum (the milk produced by the newly lambed ewe, which is high in protein, vitamins and antibodies which will initially protect the lamb). It is not absolutely vital that the lamb should have this from its own dam.

2   Check that its tail is reasonably clean and is not obviously wet and soiled from diarrhoea. The dung from a new lamb will be a thick, blackish, custard – this is quite normal and is known as meconium. The normal dung from a lamb that has received colostrum and is still wholly milk-fed can be likened in colour and consistency to a reasonably firm custard or egg yolk.

3   Put a finger in its mouth. The lamb which has fed will feel warm and may even suck. The lamb which is suffering from exhaustion, exposure and cold will be chilling to the touch and will probably offer little resistance.

4   The eye should be bright and clean.

5   The umbilical cord will still be moist in a newborn lamb but will dry to a dark shrivelled string-like appendage within a day or so. Some cords take three weeks or more to drop off, and their state is not a reliable guide to the age of the lamb.

6   The breathing should not be obviously laboured; the nostrils should be clean and free from mucus.

7   The fleece of a healthy lamb will be firm and yet springy to the touch and also very mildly greasy. It will not smell unpleasant.

Older, weaned lambs
1   Don't buy lambs with obvious pot bellies. This will possibly indicate faulty and infrequent feeding.

2   Check the wool around the tail; this should be clean and free from dung.

3   Make sure that the males have been castrated. The

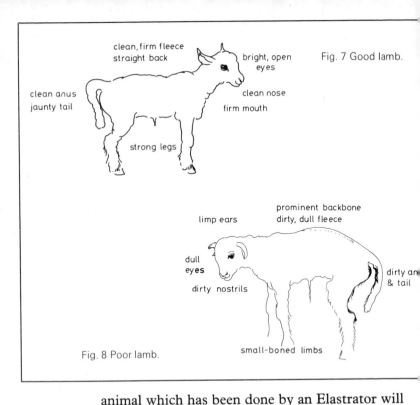

clean, firm fleece
straight back

bright, open eyes

Fig. 7 Good lamb.

clean anus
jaunty tail

clean nose

firm mouth

strong legs

prominent backbone
dirty, dull fleece

limp ears

dull eyes

dirty an & tail

dirty nostrils

Fig. 8 Poor lamb.

small-boned limbs

animal which has been done by an Elastrator will have a minute purse with no testes. The lamb which has been castrated by Burdizzo will be more difficult to judge, as the operation only crushes the tube leading from the testes. The purse however will not be so large as that of an entire ram lamb. Incidentally the entire ram lamb will grow faster and provide leaner meat than the castrated one, but he may try to mate his companions as the breeding season approaches, and will also be quite willing to fight other males in the flock at this time.

4   An animal with an alert attitude and straight back is likely to be flourishing. The humped back with head held low will show that it is not in the best of health. Look for the lamb with a good back on it, with plenty of meat on either side of the backbone.

5   Choose lambs with stocky legs and which stand
    'foursquare'. Avoid smallboned animals with
    front legs 'coming out of the same hole'.

Some common terms for different classes of sheep

Male	tup, tip, ram, heeder, haggerel, he teg (up to first shearing)
Female	ewe, theave, gimmer, she teg (up to first shearing)

Hogg, hogget, is used for both sexes.
Wether is used for a castrated male.
After the fourth shearing the sheep is full-mouthed or
aged.

Buying adult sheep
Make sure that the sheep is not obviously too heavy
for you to handle. *Horned* breeds may be caught and
cast (a method of restraint for handling—see page
121) more easily, as the horns can provide a useful grip.
    Having decided that the sheep is not too heavy for
you to manage, ask yourself:
1   Is she the right breed? If, for example you live on
    exposed, poor land the shortwool and down breeds
    may suffer.
2   If you like to spin, try to find someone who spins
    wool from this particular flock or even the
    particular individual sheep. Is the wool suitable
    for your purpose? A kempy (hairy rather than
    woolly) fleece will not knit into a fine shawl! Some
    breeds and even certain sheep within the breed are
    either coarser or finer than the average yielded by
    that breed.
3   If your main aim is the production of a good

carcase, look at the conformation of the animal. Metaphorically 'undress' her to see if she has a firm, wide body. Feel the backbone – it should be well fleshed and the tail obviously meaty, the ribs well sprung and firm with a deep, wide chest. The legs should be well set at the corners with a generous bone, with the hind leg obviously well fleshed down to the hock. Avoid the narrow sheep with scant fleshing on the backbone, narrowly set and meanly boned legs and a skinny tail.

Having satisfied yourself that she is the right type, check on visible signs of health. Perhaps the most obvious will be an alert expression with the ears held according to the particular breed characteristics. If the animal is pure bred do your homework on breed standards (study the National Sheep Association's *British Sheep*). Has she bright eyes, a clean nose and firm mouth? On inspecting the mouth, an old sheep may have several teeth missing, but if you are buying her from poorer conditions than you are going to provide (buying in draft ewes from a hill flock for example) she may be a bargain if you can buy her cheaply.

Look critically at her fleece to see that the wool is dense, bright and clean and that her body is free from obvious parasites. You can do this by opening the fleece on her flank (never on the back, where you might lay it open to rain and dust) by putting your palms together and gently but firmly sliding the hands in and then carefully opening them. The skin should look clean and firm. The wool must not show any unevenness or brittleness, which could indicate poor nutrition at some time. A well-rounded perfect ewe on the other hand may indicate that she is barren, while one who has reared a lamb or two may look a bit the worse for wear. Look at her udder, check that it is even with two good teats and feels soft and silky. Avoid the ewe with the one-sided udder, maybe with a lumpy feel to it, as this may have been due to mastitis and she may be unable to feed one lamb, let alone a pair.

Transport
Newly-born orphan lambs will travel happily in a sack wrapped round the body with the head protruding. The sack will help to absorb any urine and dung. A pair can be held on your lap in this way. Animals of any age prefer to travel reasonably snugly, and an odd lamb or two wandering loose in the back of a trailer or estate car will feel very vulnerable. *Don't* bundle them in a sack in the boot of a car—the more comfortably they travel the less stress they will suffer.

Adult sheep are best transported in a covered trailer with straw bales positioned to prevent the animals being thrown about. One adult or even two could travel reasonably in the back of an estate car if they are used to being handled. The wilder breeds and unhandled stock may well endeavour to leap through the window panes. Sheep have extremely hard heads and this, combined with possible horns, could prove damaging and even disastrous.

Settling in
On arriving home, allow the lambs particularly to rest before feeding them. The newborn lamb will require draught-free airy surroundings. Don't bring them indoors; it is far too warm, and airless conditions can lead to pneumonia, apart from the attendant mess (lambs have little or no control over either their bladder or bowels!). It is far better to build a roofed shelter, possibly from straw bales. Even the corner of a shed will do, ideally with an electric point to enable an infra-red lamp to be set up for a weak, cold lamb. Failing this I have used a rubber hot-water bottle securely wrapped in an old woolly jumper, or a stone

hot-water bottle (this will retain its heat longer). The newborn orphan is best rested for at least two hours after a journey before feeding, and then a small, warm glucose and water feed is advisable (see Chapter 6 on orphan lambs).

Older stock will naturally be frightened on arrival at their new home. Make sure that the field, fold, shelter or wherever they are to go is SHEEP-PROOF *before* you set off to fetch the animals. Get the transport as near as is humanly possible to the area in which the animals are to be kept, as a frightened sheep can dash off through almost anything; it is no respecter of either persons or property. A narrow runway made of hurdles leading to the field or shed is ideal and also has a calming effect. Don't be misguided into putting the animals straight away into deep, lush grass and clover, even if you are surrounded by it. A small fenced-off or hurdled area on a preferably short, meagre sward with a net of top quality, sweet meadow hay and a bucket of water will help the newly acquired sheep to become accustomed to its new owners and surroundings. Even better, a night and a day in a shed which is wholly stockproof with some litter, hay, water and a trough (for a handful only of oats per sheep twice a day) will prevent escape and possible hours wasted searching the countryside, not to mention the attendant risks to traffic.

Decide on your method of calling; my husband likes to whistle, and for some reason I call sheep with, 'cup, cup, cup'. Each time you look at them herald your approach with the same call accompanied by the rattle of a bucket, which they will *very* soon associate with feed. They may well stand huddled in a corner with wild eyes and poised for flight. They often grind their teeth intermittently when frightened, and this must not be confused with chewing the cud. Often they will not feed till left alone. However if they will not eat at all, a branch or two of leafy elm in the summer months will often tempt a frightened sheep. Twenty-four hours indoors will help to calm them a bit before they are moved to their permanent quarters, but remember that even a normally calm, tame sheep may *attempt* to jump out in an effort to get away when under stress.

As the animals settle you will see them lying down peacefully chewing the cud. Try to resist the temptation of going to look at them too much at first, as this may prevent them relaxing enough to graze.

It must always be remembered, though, that sheep are very conservative: new additions to the flock will need time to settle before they really begin to become acclimatized to their surroundings. They behave very like homesick children—standing around looking lost. However, most will be grazing reasonably happily even if a little fitfully after three days. Generally, the older the sheep are when bought, the longer they will take to settle, weaned lambs generally settling the quickest.

Livestock movement register
By law we are obliged to keep a record of the movements of all cloven-hoofed animals. This includes sheep, cattle, goats and pigs. The register, which must be maintained by the owners of these animals, will enable the police and veterinary officers to trace any livestock in the event of an outbreak of notifiable disease; in the case of sheep this can be anthrax, foot-and-mouth disease, or scab. An official record book may be obtained from the police, but a notebook will serve just as well as long as it contains the date and records of all stock to and from your premises, including sales, purchases, mating and slaughter journeys. The record should be completed

within thirty-six hours of the operation and must be made in either indelible pencil or ink. The police may call to inspect the book at any reasonable time.

2 Constraint and Protection

It *is* possible to tether sheep, but it has its disadvantages–the animals must be moved at least twice a day, and this can prove time-consuming when moving several sheep. The chance of the animals getting enough exercise is greatly reduced and a ewe will need to walk at least half a mile a day in order to help ensure an easy lambing. Ruminants, and especially sheep, need plenty of exercise or the chance of bloat will be increased (see Chapter 9 on health).

A very careful watch must be kept on the animals under this system, as it is very easy to underfeed them. On the other hand they can gorge themselves (especially if they are hungry) when moved from a bare patch on to an area of lush grass and clover which they might find particularly palatable. This could prove fatal under certain circumstances. However, tethering does mean that you have a high level of control over the grazing available.

I don't like to leave tethered animals unattended. Sheep are extremely vulnerable to attack by marauding dogs or even teasing children. The animals are also at the mercy of the elements, and must be placed where they can either reach shade in hot, sunny weather or be given protection from driving wind and rain. It cannot be much of a life for a sheep to be tethered; by nature they are extreme herd animals.

If at any time you must resort to this method there are one or two points that should be noted. A chain collar is most unsatisfactory, as it quickly becomes irrevocably entwined in the fleece, and if a close watch is not kept on this the metal can even work its way into the flesh. A rope collar will shrink in wet weather, and again it will become enmeshed in the wool. A thick nylon rope collar is cheap, does not shrink but it is inclined to knot the wool. I prefer a collar made of stirrup leather (made by a saddler). This can be satisfactory, but it must be inspected at least twice daily during the summer months as it can prove an ideal harbour for fly eggs which, within hours of laying, hatch into flesh-eating maggots. The leather will be naturally greased by the tallow in the fleece, but it will need to be removed and cleaned frequently to ensure that it moves freely over the wool.

Having decided, I hope, not to tether your sheep, let us look at the other methods of constraint open to us. Actually sheep are very good homers. In the mountain sheep this close association with a particular area of land is known as the hefting instinct. The ewes inculcate in their lambs their knowledge of the area, the best protected regions, the sweetest grazing areas at various times of the year, etc. Our domestic sheep will also soon learn their new home's limits. Although they will sooner or later learn the boundaries of our garden estates they may never fully appreciate that the vegetable garden is out of bounds, and that their personal pruning abilities are not welcome on the roses and many other garden shrubs. They may even sample the rhododendrons, with probably fatal results.

To maintain some semblance of a garden and to keep on reasonable terms with our neighbours some effective means of sheep restraint must be employed. Remember always that a fence is as strong as its weakest point! There are several types of permanent fencing suitable, and that of a regularly trimmed hedge probably springs first to mind. There is not one that will stand up to the depredations of sheep all the year round, and a well maintained and mature hawthorn hedge will be effective only as long as it *is* well maintained. A hungry sheep will nibble insistently at the base of the hedge at a potentially weak point, and once its head is through, the body will eventually follow; worse still, it may become inextricably enmeshed (or so it imagines), and if not discovered within a few hours it will die. Sheep easily lose hope and collapse if left unattended in a difficult situation, when they are an easy quarry for dogs who may soon develop a taste for mutton.

In order to prevent sheep gaining access to the hedge, which can in its turn afford both valuable shelter from prevailing winds, rain or snow and even shade from a relentless sun, a variety of materials can be used. Sheep netting, pig netting, electric fencing, strained wires or posts and rails are all possibilities. A combination of hedge and fence is often ideal.

Fencing can be roughly divided into permanent and temporary types. The garden farmer would be well advised to spend as much as he can afford on permanent fencing around the whole area likely to be grazed by the sheep, bearing in mind that other types of grazing animal may also be kept. The whole area, once stockproof, can then by divided by one method or another to allow the possible rotation of crops, the grazing and resting of swards or the production of hay at times of surfeit. In addition to allowing different types of stock to graze their preferred species in their own particular manner, such a system will also help to control the build-up of parasitic worms. Young lambs can also be allowed on to the best swards if creep fencing (see page 48) is used.

Unsuitable fencing
Barbed wire
This can be either permanent or temporary, but as far as I am concerned I wish it had never been invented. It can cause terrible mutilation to humans as well as to animals. Sheep seem to have a fatal fascination for it, and once enmeshed are extremely difficult to extricate. If erected too tightly it can snap with a dangerous backlash, and its tensile strength is lost if overtightened. Erected with insufficient tautness it fails to keep stock in and catches up on both humans and animals more easily.

Wooden paling
This type of fencing is often made from riven chestnut. However it is unsuitable for farm stock, including sheep, as a frightened animal can *easily* become impaled upon it in the effort of attempting to jump it with possibly fatal results.

Permanent fencing
Post and rail
This type of fencing is one of the strongest and most attractive aesthetically, but it is the most expensive both to purchase and to erect. There are several reliable firms which will quote you for anything from an oak post and rail to riven chestnut or treated deal. This is one of the longest lasting types of fence and, provided that a wire runs under the lowest rail or an extra rail is provided at the bottom to prevent creepers and lambs getting under, will prove to be elegant and trouble-free.

Posts and plain wire
This is a cheaper system, but will require sheep netting to make it wholly sheep and lamb proof. This wire fencing may have a shorter life if you live near a

polluted industrial area or close to the sea.

The initial cost of these permanent fences may prove daunting, and many are tempted to economize with cheap, untreated posts of a limited life (we used home-grown elm and alder which lasted up to ten years and then rotted at ground level). Pine posts which have been debarked and creosoted or better still pressure-treated with a wood preservative are good. Riven chestnut or even sawn chestnut posts need no preservative and last for up to forty years—we have some which we move and re-use frequently, and these are at least thirty years old. Oak makes superb posts, and a visit to a demolition contractor's yard may reveal a source, or failing this oak posts can sometimes be tracked down at farm sales and purchased quite cheaply. A short-term, home-grown permanent fencing system will, however, give you thinking time to decide what you *really* want, although by the end of its life you will be cursing the day you erected it as the sheep knock down sections of it almost unnoticed while grazing through it. Remember it takes almost the same amount of time to erect a good timber fence as a cheap short-lived one!

Concrete post and wire
This is a little cheaper than wooden posts, but concrete posts must be sunk to precisely the correct depth to achieve a pleasant, uniform finish as they cannot be trimmed to a uniform height with any ease. The wire can be left for fifteen years or so before needing replacement.

Of course you may be lucky enough to have inherited existing dry-stone walling!

Temporary fencing
This fencing will either be the complete fence or the

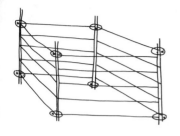

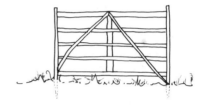

Fig. 9 These metal hurdles with interlocking connections are extremely stable. Four, we consider, are adequate for handling; six or eight would be a luxury.

Fig. 10 Wooden hurdles. These must be hammered in each time, which is time-consuming.

method of dividing your existing fenced area into smaller paddocks.

Hurdles

These can be either metal, timber or woven wattle. They may all be purchased, but the wattle and bar hurdles can be made at home. Metal hurdles are strong and extremely adaptable and can be used to make windbreaks, temporary housing for lambing, small enclosed areas for controlled grazing and gathering pens as well as races. The latter are passageways which force the sheep into single file for inspection and handling and, with the addition of shallow troughs, for foot medication. The wooden bar hurdles have sharpened end uprights which can be driven into the ground. These afford less shelter than the wattle hurdles but are easier to handle. All hurdles are about 2m long and about a metre high. The metal hurdles do *not* need to be driven into the ground in order to stay upright; they connect at their corners. A square of four hurdles is a very rigid structure.

31

Electric fencing

This is an adaptable form of constraint and can be used both as a temporary or permanent fence. The cheapest method is to use two strands of electrified wire mounted one above the other on insulators which are themselves fixed to a stake. This may not restrain lambs, and adult sheep can soon learn to creep under unless they have been well trained. The best time for introducing sheep to the electric fence is just after they have been shorn. Confine the animals to a small area and place tempting foods such as damp hay (to increase the shock), kale or grass *on to* the wires. They will receive a sharp shock through the mouth and will very quickly learn to respect the wires. It is important that the lowest wire is checked twice daily to clear any touching vegetation which can cause earthing and so reduce the shock delivered from each pulse.

A new form of electric fencing now available consists of plastic, diamond or square, woven field stock netting with plastic pointed posts incorporated into the fence at intervals. It is very light, and can easily be erected and taken down by one person. it comes in 25- and 50-metre rolls. It has the added advantage of being foxproof, which is a necessity in some areas during lambing.

Electric fencing units are often available quite cheaply at farm sales—new ones can cost anything from £30 upwards. The older units may only give a very weak pulse shock, as we have found to our cost, and the purchase of a new unit is usually worth the extra expense. These units will either work from a dry battery which will need renewing every 3 to 6 months or so, or by using a car battery which is of course rechargeable. There are also units for use with a mains supply.

Recently the Gallagher fencer unit has appeared on the English market from New Zealand. This, like a

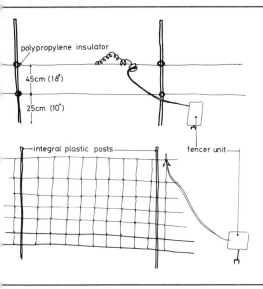

polypropylene insulator

45cm (18')

25cm (10')

Fig. 11 Electric fencing, temporary or permanent. Some sheep may need the lower wire 18 cm from the ground.

integral plastic posts

fencer unit

Fig. 12 Plastic, flexible sheep netting (good for control). Light and easily erected.

number of other types now in production, is transistorized, and so has no moving parts. Gallagher has a high energy output and maintains 5000 volts on an open circuit and drops very little voltage with a load. This effectively withers any vegetation which happens to touch the fence wire. There are several models available, including one for use on circuits up to 2km length running on dry batteries and another using a 12-volt rechargeable battery which will electrify up to 40km of fencing. There are three switch positions on each model, which determine the shock frequency. These units in conjunction with the plastic woven fencing must be as sheep-proof as anything available. I have yet to find a second-hand Gallagher, which I think speaks for itself!

When erecting any fence, remember that the closer the area resembles a perfect square the more economical it will be to enclose a particular area. Obviously the shape of the area available has to be taken into consideration but do remember that four long paddocks will need far more materials than a square one divided into four smaller squares.

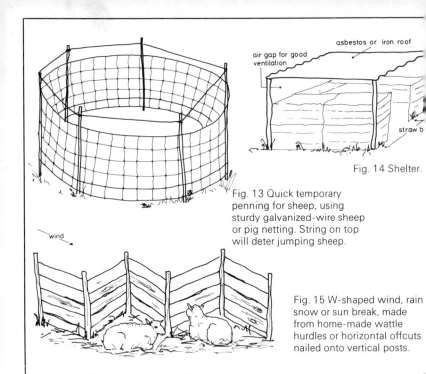

air gap for good ventilation

asbestos or iron roof

straw b

Fig. 14 Shelter.

Fig. 13 Quick temporary penning for sheep, using sturdy galvanized-wire sheep or pig netting. String on top will deter jumping sheep.

wind

Fig. 15 W-shaped wind, rain snow or sun break, made from home-made wattle hurdles or horizontal offcuts nailed onto vertical posts.

Shelter

When I think of sheep, the thick fleeced, woolly animal comes to mind. One might imagine that this coat would entirely protect the sheep from rain, snow and wind, and the mountain breeds do grow fleeces with a certain amount of kemp, which helps to shed some of the rain. This coarse kemp wool is often seen having failed to take the dye in rugs, coarse woolled socks and tweeds. But the 'improved' shortwoolled breeds and to a lesser extent the longwools whose wool is often a little coarser have had this kemp wool bred out of them over the years and are thus far more vulnerable to the rain. A thick fleece soaked through from standing in the driving rain is going to be extremely heavy for the sheep to carry around. Commercially some flocks remain out of doors all the time, although some farmers are now realizing the advantages of providing shelter during the very worst weather. We backyarders can afford to be generous and erect some sort of shelter against the elements. It will save food, as the animals will need less to keep them warm, and it will also often save the life of newborn lambs. This shelter must be draughtproof and yet airy; it must have enough room for hayracks, feed and water troughs, and space for our flock to lie peacefully without any chivvying.

A simple shelter for two or three sheep can be made from hurdles with straw bales two high for walls and corrugated asbestos or iron laid across the top to form a roof. The gap between the top of the straw bales and the roofing will ensure a reasonable movement of air, yet the bales will prevent a direct draught on to the sheep. This is mobile and easy to erect and dismantle.

A simple pole shed would be adequate—again the windproofing must not meet the roof. Small hutches just big enough for a ewe and her lambs can be erected in one corner. These can be used after the ewe has lambed to assist the mother to recognize her lamb, the mother and lambs being shut up for forty-eight hours with hay and water, by which time there will be no doubt in their minds as to who belongs to whom. Remember, a dead twin lamb is half the return from one ewe lost, while your breeding expenses will remain substantially the same.

Access to shelter will also make shepherding easier. With luck your sheep will go to the shelter on a rough night, and it will be far easier to count and check them dry and snugly chewing the cud on a stormy November evening than roaming the field with the wind and rain cutting into you, trying to find and count them.

A cheaper stopgap shelter can be made from pine or other wood offcuts erected in a zigzag. Until we erected a shelter from condemned telegraph poles and old corrugated iron our stock enjoyed a reasonably sheltered winter in this zigzag shelter. The wind never blows in all directions at one time, so there is always one side which is draught-free. On rainy or snowy nights we would take out fresh straw last thing, to find that the stock had always found the sheltered side and were lying snug and dry. We also found this zigzag shelter ideal in hot weather, as there was always shade available.

The need for shelter to the heavy in-lamb ewe will surely be obvious, as sheep love a dry spot on which to lie down. If she has to remain in the open on muddy ground she will remain standing, getting very tired. If she does eventually lie down with a very wet fleece she may easily be unable to rise again because of the weight of the fleece and the presence of mud under her, with possibly fatal results.

3 Feeding

'Good pastures make good sheep.'

Sheep kept by a garden farmer are possibly more at the mercy of their shepherd than those reared on a large commercial scale. The limited space available is going to require an active and observant eye regularly kept on the animals by everyone in the household in order that the sheep get enough to eat and also to ensure that the available pasture is used to its best advantage. Sheep, of all the domesticated farm animals, respond to careful shepherding. The expert may know all there is to know on their nutritional requirements, but unless this knowledge is backed up by careful observation by the shepherd it is useless.

The new shepherd may not know a lot about the technicalities of feeding, but if he has any feeling for stock at all he will soon know what is required, even before the sheep knows itself, by the attitude and behaviour of his charges.

Sheep, like other livestock, have an extremely accurate sense of time and will flourish on regular attention. When feeding possible extra cereals and hay during winter, before tupping (mating), and after weaning they will soon learn the regular feed times. If their meal is even half an hour late, the pattern of feeding and cudding will have been upset, and this will possibly result in less than optimum contentment and consequently growth.

Once your flock, however small, has settled it is well worth someone making a careful study of the actual feeding pattern. This is easiest in the summer, when there is often ample grass available. With adequate grazing it is usually found that by about nine in the morning they will have consumed enough grass to be lying down waiting to cud. In common with most other grazing animals and even domestic fowls, the midday period is generally marked by another rest period. Before sunset there is apparently a serious effort to stock up for the night. Of course where food is short they will spend relatively more time scouring the available area, and if really hungry will resort to bleating. The old saying, 'every baa loses a bite', is very apposite here. Having discovered what the pattern is, try to fit in your routine checks at the same time daily, so that you do not disturb them unduly. The sooner they learn to know all the family the less likely they are to be disturbed by these visits.

When the sheep are new, try to wear the same clothes each time you look at them. Back this up with your feed call. Some new sheep of ours nearly leapt the fence when I checked them one evening wearing a skirt instead of jeans. When I saw what was happening and called them, they soon realized I wasn't a stranger and that it was the usual check-up time.

Well managed lowland sheep may spend up to nine hours grazing out of the twenty-four, nearly all in daylight. Time spent waiting to cud and actual cudding may be about eleven hours, and the rest of the time will be spent just standing.

The cudding process is shared by other ruminants such as cows and goats. It enables the animals to crop quantities of mainly grass with other herbaceous plants found in pasture, swallowing this with the minimum of biting. This coarsely masticated material is stored in the rumen and soaked by liquid from the second stomach until the sheep is ready to cud. When cudding starts the food is regurgitated to the mouth and chewed repeatedly before being swallowed again and returned to the rumen. The more fibrous foods may be regurgitated several times before passing to the next stomach, where digestion takes place as in the stomach of simple-stomached animals.

There seems to be a widely held belief that sheep do not require water, maybe because they are not often seen drinking at length from a water tank like the heavily milking cow. Sheep are far smaller animals, with a relatively smaller capacity! They drink very fastidiously, with little noise. Being wary animals means that where one might notice the cow consuming vast quantities of water, the sheep may only take a sip or two before lifting its head to listen and watch.

Water is essential to life. It is one of the components of many essential body secretions. Nutrients are also carried round the body in the bloodstream, which itself mainly consists of water. It is also passed out when the animal breathes and urinates, and to a far lesser extent when it empties its bowels (the dung of the sheep is relatively dry compared with that of the cow, and in the healthy sheep is passed in the form of small, dryish pellets). The body temperature is largely adjusted by the evaporation of water from the lung surface and to a lesser extent through the skin—in hot weather sheep, like goats, may pant rather like dogs. Of course the water requirements of the milking ewe will be higher than those of the weaned lamb or empty ewe, though these will still need access to a supply of fresh, clean water.

Water containers need to be kept scrupulously clean. An old kitchen sink is often ideal, as it is easily cleaned and is large enough to allow horned animals to drink when the water level gets low. Sheep with horns coming out forwards often have difficulty getting enough to drink if the troughs are too small. Remember that during the winter, when a higher proportion of dry foods are fed, they will require more water to drink, so check that any ice is removed at least twice a day.

Where the sheep farmer can do no more than consider his sheep as a flock when organizing the feeding, the owner of a few sheep has the luxury of really observing the different needs of the individual animals. One may prefer one type of cereal and one another—you can even afford to feed the animals individually; they will well repay this personal attention with quicker growth, and you will have the satisfaction of seeing your stock contented.

As I mentioned earlier, grazing will supply the vast majority of the sheep's food requirements most of the year. Therefore the more suitable it is for their needs, the more likely they will be to thrive. Generally sheep prefer a shorter sward than cattle and will apparently derive a living off almost bare ground (but it will also be noticed that in this case the sheep spend most of their time grazing, with consequently very little cudding). Pasture that is rough and overgrown may be consumed by cattle readily, but most breeds of sheep will not like it. Controlled grazing by the use of hurdles or electric fencing will help here. The sheep are moved from one area to another according to the quantity and length of grass. The newly vacated area is allowed time to recover and regrow, while any spare areas in periods of the greatest growth (as in the late spring) can be shut off for haymaking. If the animals are to have grazing throughout the year the number retained will have to be kept down to the number which can graze adequately during the period of least growth. Remember they will need extra hay and cereals during frost and snow. Of course if they are to be housed during the worst of the winter period, perhaps one or two more sheep may be kept.

Generally, existing grass such as that found in orchards or spare areas of the garden is very seasonal in production. It may well not start growing until later in the spring than especially bred and formulated grass mixtures. It will also possibly stop growing earlier in the autumn than these purpose-bred swards. There has been a lot of work done recently on breeding for early and late grass production, but these mixtures are often not intended to be very long-lived, and also only thrive under ideal management. Indigenous garden grass mixtures may not crop heavily or long, but at least they survive, or they would not be where they are! It will often be found that our backyard sheep will improve the sward, as their grazing often encourages the finer grasses. Their tiny feet do little damage in wet weather when only small numbers are kept, and their dung is ideally dispersed in polite quantities, unlike that of the cow! Indigenous grass areas can often be improved by the use of fertilizers, such as purpose-balanced inorganic fertilizers for grass, which are sold by the large fertilizer companies. But for those who are 'organically' minded, the sward can be improved by spreading home-made compost in the autumn in conjunction with hoof and horn meal, dried blood or any other 'organic' fertilizer. A soil testing kit can be bought at a garden shop to test for any deficiencies. These slower-acting fertilizers will be available to the plant over a longer period than the inorganic fertilizers. The latter are usually applied in February or March, the grass responding fairly quickly. Animals are not generally over-keen to graze this artificially fed sward, but a heavy hay crop will often be produced this way.

A poor grazing area will often improve remarkably after a dose of lime. Your local supplier will come to test your soil and advise on the quantities needed. This can be spread over a small area broadcast by hand (if the wind is not too high), but do wear a glove. If the area to be treated is too small to warrant the interest of a commercial lime supplier, pH testing kits are available from many garden centres and will give you a rough idea of how acid the soil really is. If the land does need liming it can be applied at any time of the year, but early spring is probably the most suitable—the lime will then be soon washed in by the rain, and be readily available to the fast-growing pasture. I like to keep the animals off newly limed land for about a week after a good rain. This application will not, with luck, have to be repeated for several years. Signs of a sward in need of lime are high proportions of moss, sorrel, spurry, yellow corn marigold, dandelion, plantain, mayweed and lawn daisies in the turf.

A study of weeds on your soil will give a general indication of prevailing conditions. Waterlogged areas often support rushes, creeping buttercups, docks and tussock grass, while dry areas grow narrow-leaved grasses and generally weak vegetation with a bluish tinge. Coltsfoot will often indicate a wet clay subsoil (housing the sheep in winter may help here!). However, well-grown fat hen and couch grass often indicate a fertile soil.

Over-generous use of nitrogenous fertilizers is going to stimulate the sward growth and succulence to such a degree that the pasture will not yield an ideal sheep fodder. This sort of growth will predispose the sheep to diarrhoea. It will also grow too fast, and will possibly need topping with some sort of cutter as soon as the grass seed heads appear, to produce the short bite that sheep love.

If you decide to reseed your pasture, be guided by local seedsmen, who will know the prevailing soil and weather conditions. Some seed suppliers have ready-made mixtures available suitable for sheep; others may consider your particular conditions and make up a mixture for you. Generally prolific tillering varieties of grasses and clovers (those which send out plentiful sideshoots) are needed for the long-duration sward which will be closely grazed by sheep. These may contain mainly Timothy, Perennial Ryegrass, Red Fescue and a limited amount of wild white clover (red clover is thought to reduce the incidence of twinning).

Programme of grass management month by month

These suggestions must be modified according to the area, soil and seasonal variations.

January – order any grass seed if you are going to reseed. Resist the temptation to allow the sheep access to *all* the pasture. There is often a mild spell which produces a certain amount of growth, but this is very thin on the ground, and grazing now will put back the later spring growth.

February – try to get liming done now if the pasture

needs it. This is the time to apply compound fertilizer for an early bite (be guided by your supplier as to quantities).The pasture will benefit from harrowing, which breaks up any mats of dead grass and opens the soil to allow air, water and plant food to penetrate.

March – a compound fertilizer can be applied to any area to be saved for hay. This is the time to start the war on moles. Try the local molecatcher if you still have one, or failing this try trapping them yourself. This is the only time of year that I can catch moles, as the males are pursuing the females and their usual caution has gone to the winds.

April – the grass will (we hope) begin to grow now. Lambs will be allowed to graze forward through a creep to ensure they get clean parasite-free grazing and the best grass. Grasses appear to grow most rapidly between 13–18° C, but Cocksfoot and York-shire Fog will grow at as low a temperature as 10°C. The latter grass is often found in gardens, and although it is unproductive on a farm I find that it supplies a good early bite on one of our so-called lawns.

May – cut down bracken where present and remove it. Keep nettles and thistles cut down. The sheep may eat these later once they have wilted, as they both have a very high protein content.

June – start cutting and making hay early in this month to ensure a sweet, high-protein hay. Cut first before the grass seed heads come into pollen.

July – any new pasture seeded in April will be ready for a light grazing. This will consolidate the soil and encourage the grass plants to tiller (send out side shoots). Be ready to top any pasture which has been grazed but still shows seed heads. This will encourage more new growth (grass allowed to seed will consider it has done its job and fail to grow much more).

August – pull up any ragwort (just before it seeds),

docks and thistles. All these weeds will usually come up complete with their roots at this time of year and can be disposed of safely by burning.

September – top any pasture which again may have tried to send up seed heads; this will encourage more grass growth, which will be needed for the ewes which are to be flushed (extra feeding just before tupping). 'While the grass grows the seed starves.'

October – spread manure or compost on to any areas due to be shut up for the early bite next year. Avoid using pig slurry or manure which may contain too high a copper content (pig fattening meal contains copper), as this can prove fatal to sheep.

November and December – sheep will graze, weather permitting. Hay may be needed now. Any very wet or clay areas may benefit from the sheep being removed entirely and kept either indoors or on the driest available area to prevent 'poaching' by their hooves and consequent damage to the grass plants.

Hay

If you find during April that the supply of grass is getting ahead, i.e. the sward is obviously getting too long for the approval of sheep, then this is the time to fence off a section for hay. I have found that generally the hay made from an 'unimproved' sward such as that often found in orchards and on odd patches in gardens is best cut as early as possible, as sheep often find it unpalatable if it is left until it seeds. You may be derided by the locals for cutting at the end of May, but the protein percentage will be high and you will have good, leafy hay which, if well made, will be really attractive to sheep. In our area in Norfolk the majority of hay is cut at the end of June. The resultant hay, I feel, is only fit to be used as a fill-belly for hungry horses or cattle. Sheep are very discerning and will ignore stalky, leafless, low-value hay. Mixed clover and grass hays, lucerne hay and seeds hay (that made from the first cut of a newly seeded ley) will be found the most palatable.

Judging hay
Thrust a hand into the middle of the bale and pull out a twist of hay. No dust should fly, the stalk should bear a good proportion of leaf and the colour be greenish. Avoid any that is obviously brown or bleached and brittle apart from hay on the surface of the bale. Bleached hay throughout the bale indicates over-weathering by sun and rain. A good hay has a good 'nose' and is free from any mustiness. Bite the stalk–the taste is pleasantly 'wheaty'. However, the final arbiter is the sheep itself. When buying it can be worth taking a bale home and offering it to the sheep before committing yourself to several expensive bales.

Quantities required
Generally the lowland sheep will need more than both the intermediate and mountain breeds, in fact Cluns and Kerries (intermediate breeds) will often scorn hay until they eventually find the grass is too deeply covered in snow. The Down and lowland sheep will accept hay at any time of the year if they are not getting enough pasture.

Bales vary in weight from 12kg to over 24kg, so a spring balance will help you to calculate the average weight of the bales. Until you know the hay requirement of your particular breed and its needs (and this will vary according to the winter pasture available) you could estimate 1kg per head per day for three months per year. This should be ample, and if any is left over it will keep for the following year.

Home-made and unbaled hay is more difficult to judge. A pile 3m × 3m and 'settled' to above 2m high ought to keep four sheep for the year.

Making hay
Most of us have no haymaking equipment and the area to be cut is too small for normal farm tackle. If it is mown by hand (scythe or sickle) it can be cut a few rows a night after work and then 'made' over a longer period. You will then have the advantage of not losing it all if it pours with rain. This hay will also be in small enough quantities for anyone at home during the day to turn it all in minutes rather than hours. In good haymaking weather the hay may be turned several times in one day and will then dry and 'make' quickly. The first turn is the hardest work, as the swaths are still full of moisture. Cut the grass into rows so that between each swath there is an area of bare ground of same size and shape as that occupied by the cut grass row. This area of ground will be dried by the sun, and when the swath is turned on to it, will assist in drying the grass. The underside of the swath will now be uppermost to the sun and the ground from which the swath has been turned will dry out and warm up ready to contribute to the drying process when the swath is turned back on to it. As the grass dries into hay, toss the swaths as they are turned *quite gently*, so that the hay becomes fluffed up to allow any drying wind to assist in the removal of water vapour. Harsh turning will cause valuable leaf to break off and be lost to the hay.

If rain threatens the hay can be made rainproof by 'cocking', preferably on a tripod. Once on the tripod, if the hay is combed with a rake so that the outermost hay stems and leaves lie downwards from a central source (like a small boy's hair), rain will drain down the outside of the cock and the underlying hay will remain

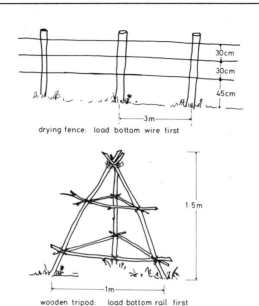

30cm
30cm
45cm

←—3m—→

drying fence: load bottom wire first

1·5m

←——1m——→

wooden tripod: load bottom rail first

Fig. 16 Hay drying.

dry and will slowly mature. The complete cock can be carted indoors by two people inserting a couple of parallel poles each about 3m long under the hay and lifting it rather in the manner of a sedan chair. Store on a bed of dry sticks topped with straw and take care not to build the stack up against the wall of the shed in which it is to be wintered, to allow air to move all round. This will help to prevent the hay becoming musty. Tread down each load and finally top off with about 150mm of straw, as this also seems to help to prevent mustiness from developing in the hay. Store bales of straw in the same manner. Do not be tempted to leave a pile of bales out of doors covered with a sheet of plastic, as these will also quickly go musty due to lack of air movement. Instead, cover with a layer of straw and then leave a gap before roofing with corrugated iron or wood. This allows any moisture to be released and applies *especially* to newly made hay.

Concentrate feeding

Many local feed mills supply ewe and lamb pellets and sometimes a rough sheep mix (coarsely ground or rolled cereals balanced with vitamims and minerals).

Flushing

This is the term used to describe the feeding of ewes before tupping (mating). The ewes are kept fairly lean until a few weeks before the tup (ram) is introduced (a fat ewe is less likely to conceive). Then say five weeks before tupping introduce concentrates rising to $\frac{1}{2}$ kg at mating time, so that the ewe enjoys a rising plane of nutrition up to conception. This can help the ewes to take quicker, and is thought to produce a higher incidence of twinning.

Winter feeding

Both lowland and Down breeds will especially benefit from winter feeding. The mountain breeds may scorn

Fig. 17 Forward lamb creep.

30cm

straw bale to stimulate interest & encourage lambs through

fresh, ungrazed grass

lamb pellets

the extra feed till later on in the winter. Try offering $\frac{1}{2}$ kg a day in two feeds, rising to $\frac{3}{4}$ kg (depending on the size and breed of sheep) after lambing. The ewe with twins or even triplets may need more. The individual care the garden shepherd can offer will help to ensure the maximum milk yield from each ewe. Do allow adequate trough space—at least 350 mm per ewe.

Creep feeding
Early lambs especially can benefit from creep feeding. At first the new lambs may try to imitate their mothers. As they gain a taste for the feed, try erecting a trough inaccessible to the ewes in a sunny, sheltered spot (but near to the ewe's feed). Early lambs born indoors can be encouraged to the creep by a lamp over the creep trough. Make sure that the feed is renewed daily and that the troughs are kept really clean. Offer clean water as well. I like to feed a small amount twice daily to prevent over-eating and the resultant en-terotoxaemia (if they have not been vaccinated against this). Some people like to have creep available ad lib. The creep-fed lamb will grow fast and will also not eat so much grass, which may be worm ridden.

Salt and other minerals
All sheep like salt, and if deprived will eat more than is good for them when they can, unless they are offered it by hand daily until they fail to come running to it. Salt can be offered as rock salt or special man-made salt blocks.

Extra minerals may be especially needed by our garden sheep where the grazing may be limited. Many feed mills or agricultural merchants incorporate a vitamin-mineral supplement in their sheep nuts and coarse sheep rations. Mineral mixes are also available for ad lib feeding. They can be put out in wooden containers whose tops are covered to prevent too much solution in the rain. The containers may be divided into two—one side for salt and the other for a general mineral mix. Sheep deprived of these minerals and salt may fail to lamb prolifically and to milk sufficiently, and their wool crop may be of poor quality. In order to be sure your sheep are getting all the minerals they need, offer them the ad lib minerals as well as that which they will be getting in the balanced sheep concentrates.

Troughs
I feel that wooden troughs are best for both cereal and mineral feeding. There are however, roofed galvanized cereal troughs available, and these can often be picked up at farm sales. The troughs are best turned upside-down as soon as the food is finished, otherwise lambs are inclined to sleep as well as dung in them. This will also prevent rain from spoiling them. I always check that all grain is finished after the animals have fed (unfinished cereal may indicate trouble). I then tip the troughs upside-down. Lamb creep troughs also need to be built so that they cannot use them as beds.

Block feeding
Within the last few years there have been feed blocks available which supply protein, vitamins and minerals in a 'waterproof' cereal base. These are produced in either cylindrical or rectangular shape and

Fig. 18 Home-made troughs.

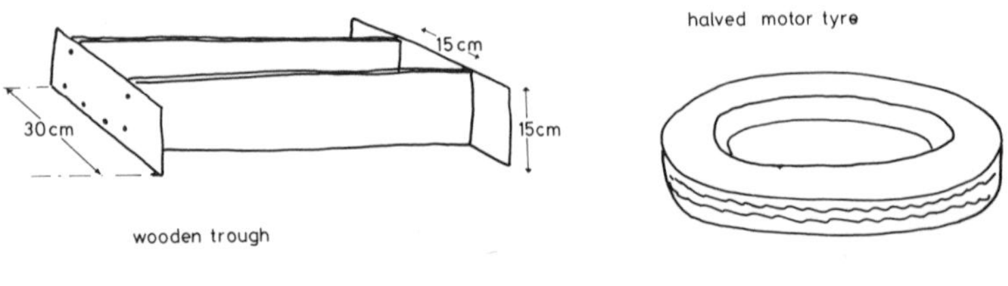

15 cm

30 cm

15cm

wooden trough

halved motor tyre

weigh about 25kg according to make. They are hard enough to prevent the animals from biting them and are claimed to be self-limiting. They are best fed during the autumn, winter and early spring. These blocks are available specifically for cattle, sheep or horses. I have found them particularly useful during hard, snowy winters as they help to balance the existing ration and also encourage cudding. The important thing is to feed sufficient roughage in the form of hay for the animals to balance their intake of the block. I give cereals as well, but the instructions only advocate this in really cold weather. One maker of feed blocks supplies a galvanized trough into which the block fits. The trough has holes at the bottom to allow water to drain away. It is important to offer the block before the animals really need it (about September according to the prevailing conditions), or you will find that the first one disappears extremely quickly. However they will soon adapt their intake to a normal quantity. Sheep will probably lick off about $\frac{1}{4}$kg per head per day depending on the other food available. One block will be needed for every dozen sheep.

Garden produce
Sheep will enjoy roots such as turnips, swedes and carrots, but these must be fed sparingly, especially in the case of the heavily in-lamb ewe, and mangels must wait until after Christmas. Cabbage leaves may be fed in moderation, just a few per head per day. Kale is best limited to barely $\frac{1}{2}$kg a day, as this causes anaemia and can be fatal. Sheep will readily over-eat it. Comfrey can be offered to sheep and is specially useful as a green food in times of drought, fed at about $\frac{1}{2}$kg per head per day during the summer. During the autumn clean sugar beet tops can be fed in moderation but again not more than $\frac{1}{2}$ kg per head per day and then *not until they are really wilted*. When feeding

spare garden produce, introduce it very slowly and keep the quantities down.

Remember sheep are mainly grass eaters, although mountain sheep have learned to come down to the villages on dustbin raids when hungry.

The secret of successful feeding of sheep lies in avoiding any sudden changes in feed, including introducing supplementary cereal feeds gradually. Continuous stocking (keeping sheep on the same land all the time) will result in sheep-sick land and a heavy worm burden; the sheep will fail to thrive, lambing numbers will fall and the wool quality will be drastically reduced, being brittle and uneven in quality.

Watch your sheep and learn from them. The old lowland shepherd would spend the morning moving hurdles and the afternoons *watching* the sheep. We can watch our sheep at any time, especially if they can be seen from the house. The contented, well-fed sheep will be very obvious, while the poorly fed one will be aggressive and determined in its search for food and will often be heard to bleat.

4 Breeding

Most breeds of sheep come on heat regularly as the days begin to shorten. The Dorset Horn however has an extended breeding season and may take the ram throughout most of the year. Generally the Down breeds will mate from August on while the hill breeds may not accept the ram till October or later. Mating ceases around Christmas time.

Once the breeding season has started the ewe will generally come on heat every seventeen days or so for a period ranging from a few hours to two or three days until she is successfully mated. It is virtually impossible to tell when she is on heat unless she is running with a ram, when, ready to be mated, she will stand still for him to mount her. Ewes on heat have been known to travel some distance to find a husband. The most effective way to get your ewes mated is either to buy your own unrelated ram (at least two years old), borrow one for six weeks or so, or even take your ewes to someone else's flock where the ram of your choice is running. Having your own ram is not really going to be practical unless you have at least five or six ewes. You will then need separate accommodation for him after the tupping period. He will also need a companion, as well as adequate feeding and exercise to keep him fit but not too fat for breeding the following season.

A good ram is going to cost a lot of money unless you obtain a proved but elderly five- or even six-year-old one. The older ram may not be so sexually active or potent, but a cheaper, young and possibly un-proved ram may not transmit his good conformation, or may even throw fewer lambs. Generally it is best to use an experienced ram on maiden ewes and ram lambs on experienced ewes.

Whatever the age of the ram he will need to be fit and hard before tupping begins. One which has been prepared for show may have been confined and fed to get him into suitable condition. This will often mean that he will be fat and may not be strong enough to stand the pace of flock mating. A leaner, more fit and active animal having had plenty of exercise and enough good-quality grazing will be able to stand the strain of the season ahead of him.

Lambing on a garden scale is best organized to

coincide with the hoped-for warm weather of spring and enough grass for the ewe to maintain an adequate supply of milk. In most areas this may not occur till about April. Earlier lambing may coincide with snow, heavy, late winter rains or March gales. As the gestation period is around twenty-one weeks this will mean putting the ram with the ewes around Guy Fawkes day.

Making up the flock
This is the term used for the selection of suitable ewes for breeding, and is generally done in the early autumn. As ewe lambs will not normally reach sexual maturity till at least six months old, unless they were early lambs and *really* well grown, it is wisest to wait until their second year to have them tupped.

Making up the flock is an important task, especially for the garden farmer. Where only a couple or so ewes are kept it will be a waste of money and time to retain less than the best of animals

Feet – inspect each ewe's feet. She must have four sound, hard hooves, these are cloven as in the diagram on page 108. See that her hooves are not overgrown, trim if they are and treat any footrot (see page 108). If she is lame she may quickly lose condition.

Udder – the maiden ewe will seldom have problems here, but check that she has two perfect teats and that the udder has soft, silky skin free from damage. In the older ewe a nick during shearing could have introduced infection. The older ewe which has previously lambed will have a larger udder, and this must be free from lumpiness and swellings, which could be the result of mastitis during the previous lactation. A soft, silky udder and even teats are essentials for a potentially trouble-free lactation.

Teeth – a ewe with some teeth missing through age or damage is obviously going to be less efficient at grazing. Some breeds are renowned for good teeth which are retained into old age, but others, often amongst the highly bred, improved breeds may lose their teeth far earlier. These ewes with some teeth missing are termed 'broken mouthed'. The prolific, meaty but broken mouthed ewe may be a bargain if she is cheap, as she will possibly flourish with individual treatment.

Reject any animals which are not thriving. These can go into the deep freeze. A tight, healthy-looking fleece, bright eye and alert attitude is what we are looking for.

Worming – it is worth collecting a few samples of dung and taking them to your vet, who will make a worm count and advise on worm treatment. This can then

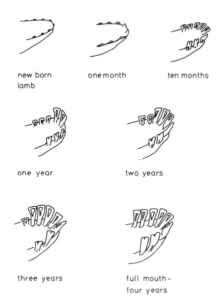

new born lamb

one month

ten months

one year

two years

three years

full mouth – four years

Fig. 19 Teeth development. The sheep only has a pad in the top jaw in place of front teeth. Upper incisors never develop. By one year a full set of teeth will have grown. At one year the milk incisors on the lower jaw are gradually replaced by permanent teeth until the time the animal is about four and a half years old, when it is known as 'full-mouthed'.

be carried out just before flushing. A move on to new grazing after worming can effectively reduce the number of parasites in the gut.

External parasites should have been destroyed during the obligatory sheep scab dip, which also takes care of lice, keds and ticks (see Chapter 9).

Flushing – while the flock is being made up the ewes are best kept on fairly poor pasture. A fat ewe is less likely to conceive than a lean one. However see that they have their minerals and water always available. A month before you introduce the ram start to feed $\frac{1}{4}$ kg of oats per ewe once a day, and if possible move them to better grazing so that the animals enjoy a rising plane of nutrition. This is known as flushing, and has been found to improve conception rates and raise the incidence of twinning. It is important for them to get enough food, but not to get *too* fat, and this is where experience helps.

Trimming – the longer-woolled ewes can have some of the wool around the tail and back legs trimmed off to assist the ram's work, but don't take off too much, as this can expose the udder to the draughts, wet and cold of winter.

The ram – some rams are never seen to serve the ewes, yet the lambs appear at the appropriate date. They often only mate at night. So that we can be sure that he is working, the ram can be marked with coloured raddle (obtained from your local supplier of animal products) mixed with grease, which is rubbed on his

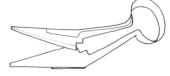

Fig. 20 Dagging shears. These are angled to ease the removal of soiled wool around the tail, which reduces the risk of fly strike. They can also be used for hand shearing though straight shears are more suitable.

Fig. 21 Holding ram to apply the raddle.

chest between the front legs. This is then transferred on to the ewe's rump as soon as she is mated. If the colour is changed after sixteen days the ewes which return to service can be noted. The ram who is accustomed to being hand-fed daily will accept the raddle being put on daily. There is a harness available which holds a week's supply of raddle but it is expensive, often slips, and may impede the ram's feeding or mating. I prefer to raddle the ram daily so that I can handle him to check his condition and also feed him if he needs it.

However, he is not likely to lose too much weight with only a few ewes to mate. It is a good idea to keep a record of the date on which each ewe is mated. This will be a big help when lambing time arrives, as you can keep a special eye open for each ewe as the day of her expected lambing comes around. With only a few ewes, you will know each one individually, but if they each have an ear tag or a number painted on their fleece the appropriate identification can be entered with the expected lambing date in the record book.

Some garden shepherds owning the less improved breeds, especially some of the so-called rare breeds, have found it possible to keep a ram in the flock of ewes permanently. The disadvantage is that you may get lambs earlier in the year than you want, and you will also have no idea of the individual lambing dates unless you keep raddling your ram from August through till Christmas! To avoid inbreeding you will have to change the ram yearly or remove all the ewe lambs at weaning. However I have seen the ram permanently retained in one particular flock of Jacob sheep (where the ewe lambs are sold each year) and in this case the lambs regularly arrived in late April and early May!

The ram is usually perfectly tame and quiet till the breeding season starts, and then you may notice him sniffing the air suspiciously. This *can* be the only indication that the breeding season has started. Do take care, as he may well become extremely bossy and be capable of jumping hitherto impossible heights to get to the ewes.

Feeding the in-lamb ewe

Pregnant ewes will of course need enough food both to maintain themselves and to allow the embryos they are carrying to develop. Generally ewes are pregnant over the winter period, when the quality and quantity of grazing is at its lowest. This will mean that extra feeding will be required if the animals are to produce healthy lambs and maintain an adequate supply of milk.

Hay – owing to the increasing size of the lambs being carried there will be progressively less room for bulky foods. However, being ruminants, sheep need a certain amount of roughage to survive. This is best supplied by feeding best-quality leafy hay ad lib in covered hay racks (or in a shelter under cover). Some of the intermediate breeds such as Clun and Kerry Hill and the mountain breeds will scorn this hay until their grazing is well covered by snow, but the majority of sheep will help themselves to increasing amounts of hay as the winter progresses.

It is well worth being very choosy over your hay. You may notice that the sheep are just not eating it, but will sniff delicately and just walk away. This may mean that it is not up to standard, and another source must be found without delay. The best hay will contain most of the minerals and vitamins needed by the pregnant ewe until about eight weeks before lambing. Depending on the breed, up to 2 kilograms (3–4 lb) may be eaten per head per day.

Roots and kale – these succulent, bulky foods can form part of the winter diet, and some even say that large quantities can be fed. However, as the pregnant ewes get heavier there is less and less room for bulk, and the succulent roots and kale, being made up of mainly water, must be fed in great moderation. Kale fed in large quantities *can* cause acute anaemia under some conditions and it is just not worth the risk. The animals rapidly lose condition, the urine may become reddish, and the white of the eye a yellow-brown. Even if the sheep do not die at first, they may not survive the actual lambing. Feed only about $\frac{1}{2}$ kg (1 lb) a day and this will supply a useful green tonic. Roots such as carrots and turnips, which again are bulky, are best fed in very small quantities of up to $\frac{1}{2}$ kg (1 lb) a day.

Fig. 22 Covered hay rack.

Concentrate feeding – depending on the ewe's condition and about eight weeks before lambing is due, start concentrate feeding. This is known as 'steaming up'. An accurate knowledge of the expected lambing dates is thus essential!

Begin with about 100 g (4 oz) per head per day in two feeds and build up to about 700 g (1½ lb) just before lambing. As the breeds vary enormously in size obviously the smaller types will need far less per day than the larger Down and longwool breeds. Be guided by the condition of the animals. The overfat animal may have difficulties, producing too fat, large lambs; the thin ewe may produce underweight, weak lambs and have barely enough milk to feed them. Check for condition by feeling the dock and along the backbone. Obviously experience is a help here. A *really* thin animal will have the bones of the back barely covered with flesh; however, the dock of a fat animal may be really well fleshed and will have a soft, fatty feel, and the bones of the back will be hardly apparent. Try to aim for a happy medium between these two extremes. Concentrates must be fed twice daily–check that they are all feeding and that the food is finished quickly. See that they have fresh, unfrozen water all the time. The concentrate ration can consist of a proprietary ewe cob or nut or a homemade mix of :

2 parts crushed oats

2 parts sugar beet pulp

1 part ground nut cake

mineral mix to be available all the time

Do remember that the rumen flora is very susceptible to violent changes in the sheep's diet, so aim to introduce the new food very gradually if you do not want digestive upsets.

Exercise

As pregnancy progresses some sheep may move about

less. The ewe needs plenty of exercise if she is to lamb easily, so it is often a good idea to arrange for the concentrate feeding to be at the furthest point from the shelter and hay. The ewe will need to walk at least a quarter mile a day to remain fit.

Observation

Look at the sheep critically as often as is convenient. As lambing approaches, increase the frequency of your visits without disturbing them unduly. Watch out for any sheep not behaving like the rest. This can be the first indication of trouble.

If your sheep is extra long-woolled around the back legs it is a good idea to trim this away from the inside of the back legs (taking great care not to damage the udder) and around the tail. This will help you to observe when she is beginning to 'bag up'. About a week before the sheep lambs her udder will begin to enlarge; by the time she is due to lamb it will be a bright, shining pink. The vulva will also become red and flabby, and if you feel either side of her tail the ligaments, which are normally tight, will be slack. The trimming will also help you to observe how the lambing is progressing without disturbing her by unnecessary handling.

Watch to see that they all have somewhere *dry* and sheltered to sleep at night. They do not appreciate a muddy bed and would rather stand. This makes them needlessly tired.

5 Lambing

Lambing kit

The provision of a lambing kit sounds ominous, but having a few basic necessities to hand may save the life of a lamb or even its mother. The following things stored in a covered box are best kept somewhere handy—you are bound to need it when you least expect it!

a) a clean coat
    (i) tied at the back
    (ii) sleeves with elastic to keep your clothes covered
    (iii) long enough to cover the tops of your boots or a long plastic apron.

b) pessaries to insert after a manual examination. These will help prevent infection.

c) non-irritant disinfectant.

d) lambing cord, a metre of thick, soft picture cord will do, boiled and kept in a clean plastic bag.

e) soap, bucket and towel to wash and dry hands.

f) bottle and teat. Buy a lamb, kid or calf teat with a large enough hole for the milk to easily drip through when the bottle is upturned. You can make the existing hole larger by inserting a thick needle which has been held in a flame for a minute or so.

g) aqueous iodine (not tincture of iodine) to put on the navel. This will dry and seal the umbilical cord and prevent infection entering the body.

h) obstetric jelly.

i) torch.

j) notebook and pencil.

Don't leave getting these things together until the expected lambing time. Aim to have your lambing kit ready at least a month before your first lambs are due.

At first you will not know how your ewes are going to respond to the actual lambing, so try to keep a note of the individual animal's behaviour so that you will know what to expect next year. There is even a chance that daughters may behave as their mothers have done. Some animals love to have moral support and will obviously enjoy your sympathetic company. Others I find will stiffen up and stop the whole process while you are anywhere near. This is where the trimmed ewe is a help—you may possibly be able to observe her (through binoculars if necessary if it's light!) and thus be ready to go to her aid if necessary.

Do remember that the *vast majority* of ewes will lamb normally without any need for assistance, and in most cases if they have not started between eleven and midnight, they will not lamb much before daybreak. An adequately fed and fit ewe will normally lamb quickly and easily on her own.

Normal lambing
The ewe about to lamb will leave her flockmates and often wander up and down the fence endeavouring to find a suitable private corner. She will then try to make a nest by scraping the ground with the front hooves, turning around till she is ready to lie down. She may bleat a little, looking around towards her tail. Normally she will lie down and very soon start to bear down. Some ewes are quite silent while others (especially first-time lambers) make a characteristic groaning bleat which will warn you that a sheep is starting to lamb. She may put her head up towards the sky and then look round at her tail, then get up and down several times trying to get comfortable. The appearance of a milky coloured mucus at the vulva may be the first indication that lambing is about to start. This mucus will have been plugging the womb during pregnancy.

The first stage of lambing

The muscular walls of the womb contract, pushing the lamb along the cervix; this is what makes the ewe feel uncomfortable. As the lamb is pushed further along, the cervix opens and the lamb enters the birth canal. The lamb pushing on the floor of the pelvis sets off the contractions of the muscle wall of the abdomen.

Second stage

Now the abdominal contractions coupled with contractions of the wall of the womb will push the lamb along the birth canal. The first things you will see in a normal presentation are the two front feet followed by the head resting on top of them, and usually a few contractions later the lamb will be born.

The birth

The 'water bag' containing yellowish amniotic fluid is sometimes broken when the ewe gets up at this stage of lambing. Don't be alarmed if the ewe licks up this fluid. She will usually lie down again as the lamb is born after probably the most intense contractions of the whole labour. If the lamb is born still in the bag, the ewe generally turns round and breaks the membrane, drinking the fluid and thus removing it

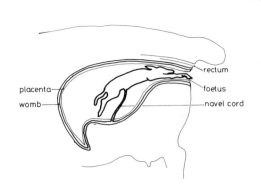

Fig. 23 Normal lambing.

from the lamb's face. In the case of an exhausted ewe this membrane must be broken and removed from the lamb's face within a couple of minutes of birth or it will die of suffocation. The actual birth will usually break the cord which has been the lifeline between the ewe and lamb for the past five months.

After the birth

Normally the ewe will stand and turn and lick her lamb. It will shake its head and often bleat while she licks it dry. This licking will help to stimulate the lamb's circulation while ewe and lamb will learn each other's smell. The lamb will soon struggle to its feet and search for the teats. It may at first look at the wrong end, but a strong lamb will soon find the udder and start to drink the colostrum with its back legs widely spaced and tail wagging vigorously. The mother will be murmuring maternal noises as she licks her young with great vigour. In the vast majority of cases the afterbirth is ejected very soon after the lamb and is generally greedily gobbled up by the ewe. Very occasionally some or all of it is retained, the first indication being portions of bloody material hanging from the vulva and the ewe possibly looking unwell. Never attempt to pull it out, but call the vet if it is still apparent forty-eight hours after birth.

Colostrum

This is made up of a concentrated collection of important nutrients laced with the ewe's protective antibodies which will help to guard the lamb from infection for its first few days. It is vital that the lamb sucks at least in the first six hours and preferably sooner. The laxative colostrum also stimulates the lamb to pass meconium, the thick, black, custard-like dung which has accumulated in the lamb's gut before it was born.

Multiple birth

If the mother has another lamb or lambs she will lie down again, and generally the subsequent births are easier and quicker—the lambs may even be delivered back legs first without difficulty. On occasions the first lamb may get forgotten, especially with a first lamber who may get up and wander off before the second is born, and subsequently ignore the first born. This is where a lambing pen will be useful. Carry the lambs (low down so that she can still see and smell them) to the pen and shut them up together with hay, water and a warm bed for a day or two to give the lambs a chance to learn their mother's smell and the mother to accustom herself to them. An elderly or tired ewe who may not take to her lamb can be shown the family dog if you have one. The instinct of any newly delivered animal is to protect its young fiercely. The lamb will be able to suck, and this often seals the bond between mother and young while she turns to face the dog.

The popular picture of the shepherd with lambs nestling in his arms may appeal, but just try walking with two lambs under your arms. Their mother will become frantically worried, desperately dashing around looking for her young. Lead the lamb with a scarf around its rump (if you have not far to go) or carry it low down at her eye level, and then she will see and so follow.

First-time lambing

The ewe lambing for the first time may well get bothered and bleat excessively. She may wander about with the water bag hanging out and the lamb half born. Try to get her to a shelter *quietly* and prevent her leaving till she has suckled her lamb. She may well be terrified of this thing dashing around her frantically butting between her legs. In most cases as

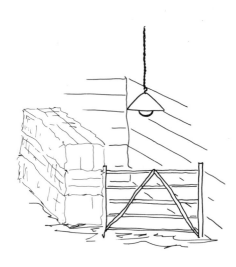

Fig. 24 Lambing pen.

soon as she feels the lamb sucking she will gradually settle and then accept it. She may back away nervously as soon as the lamb gets up, but I have noticed that as soon as she licks it the maternal instinct seems to develop rapidly. Forty-eight hours in a small pen confined together will seal the relationship. A new mother may also abandon the first of twins or allow another ewe to steal her first lamb; again the confinement in a lambing pen will prevent this happening.

Malpresentation
If the ewe is obviously straining without achieving anything there may be trouble. However a first lamber can strain for 3 to 4 hours without there necessarily being anything wrong. If you interfere too early more harm than good may be done. Give her time. Always remember that it is far better to allow the ewe to lamb on her own rather than rush in too soon. However there are certain conditions where help will usually be needed:

1 If the water bag has broken and the ewe has subsequently given up pushing.
2 If the ewe has been working hard and then suddenly everything apparently comes to a halt.
3 When you see that the presentation is obviously wrong.

It may be possible to put her in the back of a car to get her to professional help quickly. If this is not possible you will have to pick up your lambing kit and get to work. Put on your coat and boots; make sure nails are *really* short, remove any rings and then scrub up thoroughly in the bucket with non-irritant disinfectant and antiseptic soap. Wipe around the vulva of the sheep with disinfectant, then wash hands again before liberally covering them with obstetric jelly. Get someone to hold the ewe steady while you hold her tail up then introduce your more sensitive hand (your left if you are right-handed) with the fingers bunched together at the tips. DO BE GENTLE. Any internal examination is potentially dangerous to the sheep. Do not be in any hurry, work *with* the sheep and don't force her.

Examination
The first thing is to establish the type of presentation that this may be:

1 Forwards, when you will be able to feel the head.
2 Backwards, when you will feel the back legs or tail.
3 Lateral, when it comes sideways, but this is very uncommon.

The clue to a backwards or forwards presentation will be obtained if you can distinguish between back and front legs. In front legs the joints of both fetlock and knee are in the *same* direction while in the back leg fetlock and hock bend in *opposite* directions and you will also feel the tail!

The object is to get the lamb into the normal position to come down the birth canal. If it is obviously in the wrong position *wait* until the ewe is *not pushing*, then *gently* push the lamb back into the womb.

Once the lamb is back in the womb feel for the front legs of the nearest lamb if there is more than one (it is sometimes difficult to get the two front legs from the *same* lamb), put your hand around the hooves to prevent the birth canal from being scratched, and gently pull downwards in an arc following her tail, remembering to work as she strains and allowing her time to let the birth passage expand. If she has given up working you will have to pull the lamb out very gently on your own. This is where the lambing cord may be needed.

After a difficult lambing

As soon as the lamb is delivered check that the nose and throat are relatively free from mucus to allow it to breathe. If the membrane is still covering the face this must be removed as soon as possible. If the animal is not breathing grasp it by the back legs (above the hocks) swing it around and this will force the blood to the chest and start it breathing; it will also help to drain the lungs of any fluid. Place the infant lamb in front of the ewe for her to lick, and even if she is really exhausted the sight and smell of her young will stimulate her to lick and mother it. A damp, limp, apparently lifeless lamb will then shake its head and even try to bleat. Before long it will normally try to get to its feet. Now check that the ewe's milk is flowing by drawing a little from each teat. If she is too tired to stand, it's worth drawing some off into a sterile bottle, putting the teat into the lamb's mouth (holding firmly in case it is sucked off and swallowed) with the bottle upside-down. Some of the colostrum will come out of the hole, and before long the lamb will start to suck.

If the weather is really wet and windy get mother and young to a sheltered spot as soon as possible, and if necessary put the lamb under a lamp if it appears at all weak and cold. All lambs seem exhausted just after birth. Most will soon come to life, but a weak lamb quickly loses the will to live when its body warmth is removed by continual wet or high winds. A lamb which is not apparently breathing and has not responded to a vigorous swing may need heart massage by gently and rhythmically pressing the rib cage every three seconds or so. As soon as you can feel the heart beat, however weak, get the lamb into the warm. I have even wrapped a lamb in a towel and put it in the coolest of the four ovens of the Aga with the door ajar for five minutes or so to really warm it through. It will then soon get to its legs and will now need its mother's milk, so return it to her keeping and take them both to a shed for a day or so. Do remember if you have to take away a weak lamb, twins or even triplets, not to leave the strong ones with her, as she will reject the weak one when it is returned. Take them all away to be warmed and return them together when the weak one is warm and stronger.

If the ewe dies try to see that the colostrum is milked from her, preferably before she dies, as any is better than none. If another ewe has just lambed it is wise to take some from her and feed it to the orphans by bottle.

The stress of lambing can sometimes result in a pop-eyed look in the ewe. We have found that offering the animal a drop of iodine simple (Lugol's solution) on a little oats once a week will help reduce the condition. The animals only seem to eat it when they need it! This form of iodine can be obtained from some chemists.

Rearing Orphans

Orphans can be bought from neighbouring flocks to be reared at home on a bottle. When breeding your own sheep you may be unfortunate enough to lose a ewe during lambing and you will be left with her young to care for and rear either on another ewe or by hand.

Rearing orphans on another ewe

Sometimes the ewe which has lost her own lambs can be made to accept another ewe's young. Some shepherds skin the ewe's own dead lamb and put this on the orphan's back to try to deceive the foster-mother. Another method is to rub any afterbirth or amniotic fluid from the bereaved ewe on to the lamb to be fostered. A method which also works is to rub a little Vick or Elliman's embrocation on to the ewe's nose and also on the lamb's shoulders and around its tail. She will then have difficulty in smelling the strange lamb's scent and may be less likely to reject it so strongly.

It is best to confine the ewe and lamb together fairly closely. The ewe can even be restrained by her head in a yoke with food and water available: the lamb will then have less difficulty getting to the udder. Some ewes kick very vigorously. In this case tie the ewe's legs above the hocks to enable the lamb easily to get to the udder. The lamb may need to have milk squirted on to its nose to encourage it to drink at first, especially if it is at all weak or depressed by constant rejection. A little warm milk from a bottle may give the lamb the initial energy to persist in its effort to get to the udder. The ewe's legs can be untied as soon as the lamb shows the persistence which will ensure that it gets enough milk even if it still gets kicked a little.

If efforts to unite ewe and lamb fail, the introduction of a dog (just the other side of the gate) will often work wonders. The ewe will naturally face the dog with her head lowered, and this will present the opportunity for the lamb to suck while her attention is distracted. Generally as soon as the lamb has relieved the ewe's full udder the mothering instinct is awakened and before long the ewe will be turning around to clean the lamb's tail as if it were her own.

Rearing bought-in orphans

When taking on other people's orphan lambs do try to see that the lamb has had some ewe's colostrum, even if it was not its own mother's. A good test to see if the lamb has sucked is to put a finger in its mouth. The lamb which has already had a feed will usually suck strongly with a warm tongue. Suspect the lamb which has little sucking instinct and a cold tongue. I have taken nearly a week getting a lamb like this to suck. The presence of meconium on its anus will often mean that it has had a drink of colostrum.

Freezing colostrum

If you are in the fortunate position of having sheep or even cattle or goats, it is always worth freezing some of their colostrum, milking it as soon as possible after birth, preferably before six hours have elapsed. Milk it directly (with very clean hands) into sterile bottles or jars. I like to have it in about 50–100 g quantities. Seal, cool under cold running water, and then deep-freeze as soon as possible. This can be thawed as needed and warmed by standing the container in hot water. Do not boil, as it will curdle immediately. This colostrum will contain some of the antibodies needed by the animal to protect it against infections on your premises. The colostrum will not change to normal milk until about the fourth or fifth day.

If you have no colostrum available it is possible to mix a milk feed which should help the young lamb. Using a quarter litre of ewe-milk replacer (made up to maker's instructions) mix in a dessert spoonful of real honey (not blended honey spread), a small or bantam's egg, and a couple of dessert spoonsful of *fresh* cod liver oil. Divide this into four feeds, having really shaken it well. Heat each quantity to blood heat as needed in a bottle standing in a bowl of warm water. After this, gradually change to pure, warm ewe-milk replacer or whatever you are going to use.

Cow's milk preferably from high-butter-fat cows can be used, although some people do experience difficulties with it. Lambs seem to have far more difficulty digesting it than ewe-replacer milk or that from goats. They seem to be more liable to scour. It has been found that to get any degree of success with cow's milk it must be boiled before cooling to blood heat. Lambs reared on cow's milk are sometimes seen to thrive for three weeks or so and then suddenly die, but on the other hand some people I know have great success with it.

Rearing lambs on substitute milk
A trouble-free substitute, though not necessarily the cheapest, is a ewe-milk replacer. This can be obtained from most feed mills and some agricultural merchants. It is more foolproof to use than the cheaper dried milk which is sold for calves. Goat's milk is a good substitute. You can even use your own goat colostrum for lambs, as this will contain the antibodies which have been built up on your own premises. But don't use skim milk of any kind, or lambs will probably scour and become pot-bellied and fail to thrive.

Quantities of feed
Most hand-reared orphan lambs who die have

mountain & small breeds: up to 2 pints daily
larger breeds: " " 4 " "

first ten days: 4 – 5 feeds daily
up to 3 weeks: 4 " "
up to 6 weeks: 3 " "
thereafter 2 " " or as required

Fig. 25 Bottle feeding.

succumbed to over- rather than under-feeding. I
know that twice a day cold milk feeding is practised on
some establishments, but I still feel that a weak
orphan is going to benefit from frequent, very small,
warm feeds for the first few days at least.

It is impossible to be exact as to the quantities
needed, as lambs vary so much in size both between
and within breeds. Aim for the lamb's sides behind
the ribs to be *flat* at the end of a feed. Any suggestion
of the tummy becoming rounded with milk must be
avoided at all costs. It is best for the lamb to look lean
and empty and still be asking for more at the end of a
feed. Most lambs are greedy suckers once they learn
the technique of sucking from a bottle, and will bleat
piteously for more. Too much milk will soon build up
into an undigested lump of curd in the stomach,
which will cause digestive upsets and pot bellies.

Normally strong lambs will be happy with five
four-hourly feeds at 6 am, 10 am, 2 pm, 6 pm, and
10 pm. This can be reduced to four times daily at a
week old and three times a day by three weeks,
depending upon the animal's condition.

Lambs which show little inclination to suck will
have to be fed by teaspoon. I put a teaspoon in the side
of a the lamb's mouth and stroke its throat at the same

time. This stroking stimulates the swallowing mechanism. Most lambs which have not sucked at first will respond by sucking a warm finger dipped in milk as they become stronger. These weaker lambs may need a 2 am feed for the first few days. I find that it is often worth a broken night to build up an ailing or small animal.

The first feed given to an orphan lamb, especially if it has travelled any distance, is best given a couple of hours after arrival to allow the lamb to settle. I like to offer a glucose and water feed – a teaspoonful of glucose to 50 g (50 cc) of water in a bottle and then another four hours later. For the following two feeds give milk diluted with equal parts of water and gradually over the next two days build this feed up to full concentration (according to maker's instructions if using powdered ewe-replacer milk).

When using fresh goat's milk I used always to boil it, but I find I get better results using *fresh* milk just warmed to blood heat. Do always be sure that the bottle and teat are sterilized either by boiling or using a dairy sterilant or Milton between feeds.

Housing
Our first lambs turned up wet and cold in a box on the back doorstep. Being overgenerous we kept them indoors beside the Aga until the floor was swimming and the whole house smelt of sheep. The lambs also developed the happy knack of finding every open door. They now go straight out to a straw bale kennel and run, large enough to gambol about in, in the straw barn. There is an infra-red lamp if they feel at all cold. Failing this you can use a well wrapped-up hot-water bottle. Fresh, draught-free air and sunlight are vital.

Lambs are remarkably agile when well fed, so make sure they have no steps up the straw made by shifting bales. Wooden partitions are ideal here. Keep the litter clean by adding enough straw daily to maintain a warm, clean, dry bed.

Solid feeding
I like to get the lambs growing really well by offering a few ewe-and-lamb pellets after a week. Place one or two in some milk, but before they go soft push them one by one into the lamb's mouth. Within a day or so they will look for these and a few can be left in a trough. Do be sure to renew them at each feed, as stale nuts are not palatable.

A small handful of hay can be tied on to the wall of the lamb house and renewed twice a day. It is surprising how soon the strong lamb will start eating it, and this will encourage the rumen to develop with resultant cudding. Fresh, clean water must always be available of course.

Grazing
It is obviously unsuitable to put new lambs out to graze in wet, frosty or snowy weather. If there is a sunny day I like to take the lambs out for a while, making sure there is a windbreak for protection. This will also act as a hide behind which you retire as you leave them! One year when we had only one lamb we found that if we walked out to the cows the lamb would follow. It would then transfer its attention to the cow as long as we crept away on all fours. We seemed to be unrecognizable as the human feeders when we reduced to a quarter of our normal height, and so we solved the problem of leaving a lonely, bleating lamb. It had transferred its allegiance to the cow!

By the time the lambs are six weeks old they can be reduced to a morning and evening bottle or even just an evening one; then as they reach two months the milk feed can be cut out. Depending on the quality and quantity of grass the concentrate ration can also be stopped. If the lambs seem contented, spending much of the time either grazing or cudding, it can be assumed that they are being adequately fed. By three months old the best lambs may be ready for slaughter.

Dung

A fairly accurate judgment of lamb health can be made from the state of the dung. The newborn lamb will pass a dark, sticky material, known as meconium. As the colostrum fades to milk this changes colour to yellow (darker or lighter depending on the type of milk the lamb is receiving, i.e. ewe's, ewe-replacer, goat's or cow's milk). It is generally a well-formed, sausage shape which normally drops to the ground leaving the anus clean. If the lamb has any difficulty passing dung a very small quantity ($\frac{1}{2}$ teaspoon) of brown sugar can be added to the milk until the correct consistency of dung is passed. As the lamb starts to eat solid food the dung will change to pellets.

If the dung becomes loose and even watery, leave out the next feed, substituting for it a bottle of warm water with a teaspoon of glucose added if you feel the lamb is at all weak. The following bottle is best diluted half water and half milk, but keep up the correct *quantity* or the animal can quickly become dehydrated. BCK (bismuth, charcoal and kaolin) granules can be obtained from your vet if you feel that the lamb's looseness is 'mechanical' rather than caused by an infection. This is best mixed in really hot water to break down the granules and added to the milk feed according to quantity and instructions on the container. Naturally the dung will change to a charcoal colour until it has all passed through.

As the lamb consumes more nuts, hay and grass the dung will change into the pelleted form of the adult sheep. Sudden changes in routine, temperature and diet can all affect the digestion, and this is first seen in the dung, so try to make all changes as gradual as possible. It can take up to three days for any alteration in management to be reflected in the dung.

7 Meat

Meat obtained from home-grown lamb can taste far sweeter and more tender than any shop-bought lamb.

It may be possible to kill your best lambs (if not needed for replacement) any time from three months of age, depending on their condition and breed, although later born lambs may pay for keeping over the winter if you have enough grazing and winter fodder available.

There is nothing to stop you eating an older sheep, for instance a ewe which has been culled. Her relatively unstressed life in the small flock will usually ensure an excellent flavoured mutton. The barren ewe is often fat, so it may be worth waiting until the winter to kill her, by which time she might have slimmed down a little.

Judging fitness of a live sheep or lamb for slaughter will need a certain amount of experience. The first lamb we killed was allowed to grow far too large and fat merely because we were unaware of the method of judging either weight or fatness. If you pay a visit to your local weekly stock market you can watch the dealers as they handle the fatstock to be sold for slaughter. After handling a few from each pen you will soon get a reasonable idea of the feel of a fat sheep. Generally the weight of the butcher's carcase will be around half the live weight.

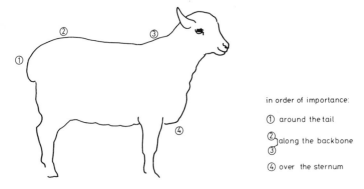

in order of importance:

① around the tail

② along the backbone
③

④ over the sternum

Fig. 26 Handling points to assess fatness of sheep.

Fatness – this can be judged by feeling the dock (tail), over the loins, down the backbone and across the shoulders. The individual bones of the dock will be clearly felt in the thin animal, but in the fat one the layer of subcutaneous fat will almost completely cover the bones. The ideal state is about midway between these two extremes. The individual bones of the vertebrae will be prominent in the thin animal and fleshing will fall away sharply on either side. In the fat lamb the flesh will almost completely cover the vertebrae and will lie horizontally on either side of the back, although this depends upon the breed to a certain extent, the lowland breeds usually being meatier than the hill ones. If you open the wool on the animal's side you will be aware of a certain creaminess in the skin of the fat animal while the colour will be a deeper pink in the thin one. However for a quick judgment I like to run my fingers around the dock.

Conformation – of course this will depend on the breeding of your stock. No amount of good feeding will improve the basic shape and capacity for fleshing. Many of the relatively unimproved breeds, Soay, St Kilda, Jacob, etc., *may* have comparatively poor fleshing on the leg and loins compared with the Down breeds.

Slaughter

This can be carried out by your local slaughterhouse. We originally tried several abattoirs and have now settled with one which is small and friendly. We contact the owner the week before we intend killing to inquire which day he would prefer us to bring the animal.

Transport – sheep, being relatively light, are a little easier to transport than larger animals. However a trailer is really ideal, although the back of a Land Rover or even an estate car is possible. Handle the animal with great care to avoid bruising, and pack it in reasonably snugly with a bale of straw if necessary to prevent it dashing about. Some slaughterhouses like the animals there the night before in order for the gut to be reasonably empty before slaughter (food takes up to four to five days to complete its passage through the sheep, so yours will not suffer from one night's starvation!). Water and bedding will be provided. The sheep will also be reasonably calm before slaughter, which will help to improve the flavour. A worried animal will often kill out tough.

The slaughterhouse – killing is usually carried out early in the day, so if you want to take home the skin, head and pluck (liver, lights, heart and sweetbread) find out what time it is convenient to collect them (usually before midday). The meat inspector calls daily to check that the carcase is fit for human consumption, so wait until he has called. Inspected meat will be indelibly stamped on both sides of the carcase, which is not necessarily wholly condemned for one fault. One of our lambs was passed except for a leg which was rejected because of an infection; in this case the leg was cut off and destroyed and we were left with a three-legged lamb. The carcase is best 'hung' for about three days, and most slaughterhouses will mature your carcase in their cold store.

The pluck, head and skin

The *pluck* is best taken home on the day of slaughter, the liver being sweetest when eaten as soon as possible. The *head* is often not skinned unless you especially ask for it to be done. It is a rather grisly task, but the resultant head makes a most superb broth. Cut away the skin with a really sharp knife, starting at the neck end, snip off the ears and then steadily chip the skin off down to the nose and mouth parts. If you like the brains, chop the head longitudinally down the skull and midway between the eyes (you may not be surprised by the smallness of this organ!). Remove the eyes using a pointed knife to cut the muscles behind them. Wash the skinned head, put to simmer for several hours with an onion stuck with a few cloves and a couple of carrots, a stick of celery and a handful of pearl barley. When the cheek flesh falls away easily, remove the head and pick off the meat from the outside and inside of the skull bones, chop in neat dice, season and serve with the liquor, vegetables and pearl barley.

The *pluck* can be made into haggis. Use only half the liver. Place half of the liver, the lungs, heart and windpipe in a saucepan of boiling water with a teaspoon of salt and simmer for two hours. Mince all the meat (except for the windpipe, which can be used for dog or cat food) and mix with a $\frac{1}{4}$ kg of minced suet and $\frac{1}{4}$ kg of medium pinhead oatmeal (it must be pinhead to make a good haggis). The oatmeal must have previously been toasted to a golden brown in a medium oven. Add one tablespoonful of finely chopped, minced herbs, four chopped, medium-sized onions, half a teaspoon of pepper, one teaspoon of salt, and a small sherry glass of whisky (this last may be omitted in extremis!). Put into three pudding bowls, cover with greaseproof and steam for three hours. Each pudding will serve five to six people. This freezes well. Thaw overnight and boil for three hours the day they are to be eaten. The traditional method is to wash the sheep's stomach and use this instead of bowls!

The *skin* must be cured as soon as possible after slaughter. A lamb will have little fat, but on an older sheep this should be removed from the skin as well as any flesh. I do this while washing it in rainwater to remove any blood. Leave the skin to soak in a container of rainwater for twenty-four hours, changing the water twice. Make up a solution of $\frac{1}{2}$ kg of alum (obtainable from a chemist) and $\frac{1}{2}$ kg of salt, using hot water to dissolve them both. Add about 20 litres of water and immerse the skin completely. It does not matter if the resultant mixture is warm. Leave for four to five days, stirring it each time you think about it. Take the skin out and spin-dry (without rinsing) to remove the moisture, then apply a mixture of one part egg yolk and three parts warm water (two egg yolks will probably be sufficient) to the skin side, rubbing it in well. Fold up skin side inside for about three days and hang up in an airy shed to dry. Each time you pass give the skin a wriggle and the leather will appear under your fingers as it dries. I find this a foolproof method. In warm weather the whole process need only take a fortnight!

After the *carcase* has been hung it is best butchered as soon as possible after collecting it from the slaughterhouse and while it is still chilled. We watched a butcher friend cut up several lambs, and now I feel able to tackle the job myself, although it may not be very neat. You need a good butcher's knife, a chopper and a smaller knife. It is worth investing in

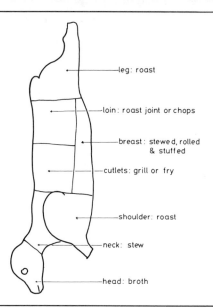

leg: roast

loin: roast joint or chops

breast: stewed, rolled & stuffed

cutlets: grill or fry

shoulder: roast

neck: stew

head: broth

Fig. 27 Jointing the carcase.

knives from a butcher's sundriesman, which will last you a lifetime once purchased. Freeze the joints after bagging and labelling.

Home slaughter – if you are eating the sheep yourself you may slaughter at home. However it will not be inspected. Pen the animal to be slaughtered overnight with no food but plenty of fresh water. Be sure it has not been dipped for at least two weeks before, or had any medication or injections. One treatment for frothy bloat leaves the meat tainted with peppermint flavour for some time! Home-reared lambs will be tame so will follow their feeder to a trough of food, where they can be shot with a 0.22 rifle between the eyes (firearms certificate essential).

Mutton recipes

It is often difficult or impossible to obtain mutton at the butcher's nowadays, so there seem few recipes for it in the modern cookery books. I will not include lamb recipes as these abound!

Mutton has a far more mature flavour than lamb.

The joints are also larger and will need longer, slower cooking.

Roasting
Season the joint as if for lamb, but allow 70 minutes per kilogram at 175°C (350°F). If you have a solid fuel oven on all the time, mutton is greatly improved by cooking slowly in the cooler oven having given it an initial five minutes in the hot oven. In this case I allow eight to twelve hours in the simmering oven.

Boiled shoulder of mutton with onions
Put the shoulder of mutton in a fish kettle with water and a little salt and *boil gently* for two hours; it should be done by then. Put on a dish and cover with the following sauce: chop six onions and boil 20 minutes in water, when the water will be reduced to half pint; add 50 g butter, $\frac{1}{2}$ litre of milk, pepper and salt to taste; stir and boil for ten minutes, then pour over the shoulder.

Sheep weight for age
Live weights will vary according to the feeding and breed of the animal. The large lowlands are killed early and smaller breeds retained longer.

Age in Months	Large Lowland	Medium Breeds	Hill and Small Breeds
0	4.5 kg (10 lb)	3.5 kg (8 lb)	2.5 kg (5 lb)
1	13	8	5
2	22	14	9
3	30	20	12
4	40	25	17
5		30	20
6		34	24
7		39	28
8		43	31
9		47	36
12			45

Using Your Wool

The only person capable of producing a really beautiful finished garment made of wool, whether it is woven, knitted or crocheted, is the one who owns his own sheep. Breon O'Casey in a recent article in *Crafts* magazine says: 'The first step to making a good rug is to buy a flock of sheep. The second step is to tend and shear them, the third step is to sort and spin the wool, the fourth is to dye it and only the fifth step is to weave it into shape. And each of these steps is of equal importance to the excellence of the finished rug. So you do all of that and then you'll see I'm right: the actual weaving is easy.' What he is saying is that only then can you fully understand and appreciate the material you are working with. Wool is the easiest of all fibres to spin and weave; each fibre has little barbs on it, so fibres will lock together and trap air to form a warm, strong thread. Before the industrial revolution, and the production of man-made fibres, sheep were bred in England for their wool, and incidentally for their milk. People spun and wove themselves garments of wool and were much healthier and, I suggest, happier for it. For someone who already has the sheep and tends and shears them himself, there are only three more steps to go: sorting and spinning the wool, dyeing it and weaving or knitting it.

To give some idea of the vast range of wools in England here are some facts about the various qualities and their different uses.

Breeds

Mountain breeds
These are sheep of hills over 300 m. They are small and very hardy, with strong, coarse fleece with a medium to long staple which is often kempy. (Kemp

is the coarse hair that sheep grow in wet areas. It tends to make the end-product scratchy and does not take dye.) It is unlikely that you would have any of these breeds in the backyard because they thrive on the mountains and need to roam, but you might be given the wool to spin. Scots Blackface, Herdwick and Swaledale all have long, coarse fleece most suitable for carpets. Cheviot and Cotswold have crisp, white wool of medium staple which is usually good for beginners and is used for tweeds and woollens. Exmoor Horn and Welsh Mountain both have soft wool with good felting qualities which is good for fabrics, while the Lonk has full, soft fleece for tweeds and blankets. In a class of its own is the beautiful soft Shetland wool with all its variety of natural colours.

Longwools
These are breeds of richer more grassy lowlands. They are bigger sheep with long lustrous wool that is best spun worsted fashion. The staple can be anything up to 670 mm in the Lincoln. Leicester, Border Leicester, Devon Longwool, Kent, Romney Marsh and Masham are all longwools; some, like Masham and Romney Marsh, are easier to spin as they are less silky and not as long as the others.

Down breeds
The Southdown, Hampshire, Dorset, Wiltshire, Suffolk, Shropshire and many others make up this group. They produce mostly close, fine wool of short or medium staple with quite a lot of crimp. On the whole if not too short, they are easy to spin.

Mountain breeds produce coarser, kempy wool more suitable for coarse tweeds and carpets. The lowland sheep produce silky, lustrous wool best turned into fine, silky fabrics, while the Down breeds yield soft, crimpy wool good for knitting and soft woollens. This is a very general picture, and of course amongst them are all the myriad crossbreds and the rare breeds. Jacob sheep are very suitable for spinning, and have the added advantage of giving two or three colours from the same fleece. As will be realized from the various divisions, the wool is affected by the conditions in which the sheep lives. For example a sheep kept in wet, windy, rough pasture will produce a different fleece from one of the same breed on a sheltered, grassy sward.

All of these wools are fascinating and easy to spin, but as a beginner it is wise not to try your patience too much. To start with, use a fleece of medium staple, that is 75–100 mm in length, which is not too silky or too crimped, even if you have to buy half a kilogram from a local spinning shop. Once you have got the knack of spinning then you can spin anything – angora, cotton, flax, silk, even nettles (which produce a cloth called 'ramie', used extensively in England in earlier times).

Sorting

When you have chosen your sheep, tended it and shorn it, you will find yourself with a large bundle of wool, the size depending on the breed of sheep. A Jacob for example yields about 2 kg of wool, a Shetland 1 kg and a Lincoln Longwool up to 9 kg or more. You lay it out on the floor like a sheepskin rug and then take away the worst wool first – the britch – then work inwards to the best, which is usually called First Diamond.

In some fleeces the quality varies very little, while in others it is quite considerable. Sorting is important, as the texture and shrinking qualities vary. Sheep have a habit of standing with their backs to the bad weather, which accounts for the Prime and Second

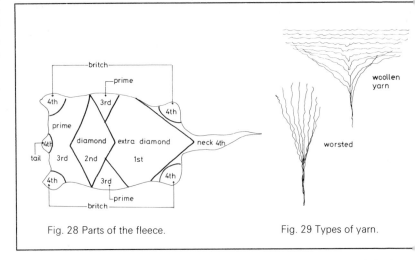

Fig. 28 Parts of the fleece.

Fig. 29 Types of yarn.

Diamond being the less greasy and coarser wool. It is at this stage that you decide what you are going to make with the different qualities. The softest wool is best for knitting and the coarser for rugs. There should be no need for any waste, as even the worst wool, if it cannot be spun, can be used for stuffing things; or you could learn to make felt.

You also decide whether or not to wash the wool. The advantage of spinning 'in the grease', as it is called, is that your hands are softened by the natural lanolin and you do not have to add any natural oil. Also washing fleece is more of a bother than doing the same with hanks of yarn. Spinning such a fleece is best done when it is freshly shorn, but the disadvantage is that if the fleece is soiled badly much of the dirt will be 'spun-in' and subsequent washing will never remove it all. If the fleece is very greasy and dirty and has been a long time shorn, spinning will be very difficult. If you wish to dye at this stage a preliminary wash is necessary. Weather conditions also affect the fleece – in very cold weather you will probably have to warm the fleece up, because the lanolin will become congealed and hard; in very dry weather you may have to add moisture or oil.

Lanolin is extracted at the washing stage, but this is usually done when large quantities of wool are washed and then the water is chemically treated, or the wool is put through a centrifuge to extract the lanolin. Before it can be used the lanolin must be purified and refined, and this is really beyond the resources of the small-scale sheep keeper.

Washing
If you decide to wash first, proceed as follows for both fleece or wool hanks:

1 Dissolve pure soap or a detergent designed especially for wool in hand-hot water. If you want to remove the dirt but keep the grease, you can try washing in cold, soapy water, but make sure that you dissolve the soap first.
2 Immerse the wool with enough room for it to soak and leave for at least two hours or preferably overnight; *do not handle.*
3 Pour off the water carefully; the best way is through a colander.
4 Rinse in the same way in water at the same temperature as that from which the wool has come. *Extremes of temperature and excessive handling felt the wool.*
5 Spin-dry in a pillow case or hang in a net bag in a windy place until dry.

Well washed and dried wool will have to be oiled prior to being spun. A little olive oil rubbed on the hands while spinning or a solution of three parts by volume of oil, two parts water to one part ammonia (to produce an emulsion) dropped sparingly onto the wool while you are teasing it will be adequate.

Teasing and carding
If you have a perfectly clean fleece of medium staple it

can be spun straight from the fleece. In order to make a WOOLLEN yarn you pull the wool out *across* the staple. This is the most usual method of spinning. If you pull it out the *length* of the staple you are spinning WORSTED. This is more usually done with a longer stapled wool which is combed first.

Teasing

Most fleece has to be teased (the word comes from the teasel plant, *Dipsacus fullonum*, whose fruiting head was originally used for the job). This means first pulling the locks gently apart, so as not to break the fibres, and shaking them as you go to remove burrs and bits of dirt which have become embedded in the fleece during the year. At this stage the fleece can give the observant much information. An illness or poor diet at any stage in the year will show as a weakness in the wool; if there are too many second cuts in the fleece your shearing technique needs to improve, and if bits of next year's growth have to be removed your shearing date was wrong. If you got the fleece covered in straw when you were shearing you will swear never to do it again after you have picked it all out! You should finish with a foamy mass of wool from which you can either spin direct or further prepare by carding.

Carding

Carding is a combing process which results in a roll or rolag of wool which looks like a magnified piece of yarn and which considerably speeds up the spinning process. It takes time and patience but it is very important to the end result. Carders are supplied by spinning and weaving shops. They are expensive, but in the long run are worth it; they should last for ever, especially if you buy the leather ones.

Both carding and wool handling can be hard work,

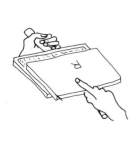

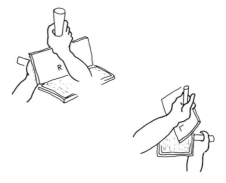

Fig. 30 Carders, and the correct way to hold them.

Fig. 31 Carding. *Above* – transferring right to left; *below* – transferring left to right.

particularly for the beginner who is using arm and finger muscles whose existence were hitherto unsuspected, and is probably also working much harder than is really necessary. The best and really only way to learn these techniques is to ask someone who is in practice to show you for a few moments. One of the best ways of learning the crafts is to join your local Guild of Spinners, Weavers and Dyers. Alternatively most equipment suppliers have information on suitable classes.

Spinning

The best way to learn to spin is on a hand spindle. This gives you a chance to learn to handle wool at reasonable speed. Once you have mastered this you can go on to a spinning wheel and to fibres such as flax, cotton, silk, angora and mohair. Any stick with a whorl will do; you can make one or they are quite cheap to buy. Tie a length (about 1 m) of heavy woollen yarn to the spindle, twist it up the spindle and make a half hitch at the top. Take a handful of the teased wool in the right hand and hold the yarn in the left hand first finger and thumb with the spindle dangling below. Allow about 130 mm of this yarn to lie over the teased wool in the right hand. With the left hand helping, draw out a few fibres to wrap around the yarn.

Drafting – with the left hand start spinning the spindle in a clockwise direction and quickly move the left hand up to help the right hand *drafting* (drawing out) the fibres. Close the thumb and first finger of the left hand on the wool to prevent the twist going up into the yarn before it is properly drawn out. (If by this time the spindle is starting to revolve backwards, put in on the floor and hold it with your foot.) Draw out about 75 mm of the fibres and then release the thumb and first finger of the left hand. Move hands up, spin the spindle and repeat the process. The art is to draw out an *even* amount of fibre each time so as to get a continuous yarn. Do not worry if you do not at first – it is more important to get your hands working correctly.

If you are left-handed or find it easier, reverse the hands in the above account.

Winding on – when the spindle reaches the ground you have to wind the yarn onto the spindle. Release the half hitch, wind on up and down the spindle leaving yourself enough to hitch on, join and start again. It is important to keep a hold on the end of the yarn with the right hand or it will unspin.

Increasing the draft – when your hands get used to their work you will probably be able to increase the amount of wool you draw out while the spindle is spinning, and this will produce a less tightly spun yarn. Yarn spun in a clockwise direction is 'Z' spun, anti-clockwise is 'S' spun.

Taking off the yarn – wind off the yarn into a skein. If you can, push the whole cone off the top of the spindle. Traditionally a niddy-noddy is used, but anything that will stretch the yarn into a long hank will do – like the back of a chair. Tie the skein in four places, because whatever you wish to do with the yarn it has to be washed now to fix the spin, and the tying, if correctly done, will prevent tangling.

The spinning wheel
Once you have got the hang of the spindle you will find you want to produce a lot of yarn, and this is where a spinning wheel will help.

Do not buy anything but a new wheel without *expert* advice. Old wheels suffer from complaints like 'wheel wobble', which makes the driving band come off constantly, and this is very frustrating. On the other hand the wheel may have become worn, and so the tension may slip. It can have been put together by an antique dealer with no knowledge of spinning, and the parts may not match. It may even be suffering from woodworm.

The spinning wheel is a delicate instrument requiring craftsmanship to make it work properly, and it deserves to be well cared for and regularly maintained. Oil should go on the axle and the foot treadle bar and where the bobbin rotates. The tension should always be slackened when not in use. Very good wheels can be obtained in kit form.

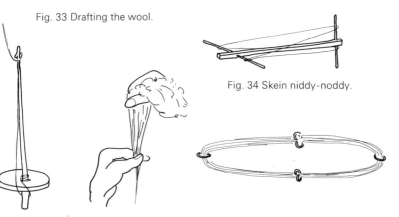

Fig. 33 Drafting the wool.

Fig. 34 Skein niddy-noddy.

Fig. 32 Tying the spindle.

Fig. 35 Tying a wool skein.

Some wheels have only a single driving band, and the tension is created by a braking device on the bobbin called 'Scotch tension'.

Spinning on the wheel is just the same as with the spindle except that the twisting of the spindle is now done by foot treadling. This is the first thing to practise on the wheel, so that you can forget about it and concentrate on your hands. A good way to practise is to ply, which is done by putting the balls or bobbins of wool behind you, in a saucepan to prevent them from rolling all over the place, and then bring the wool over the back of a chair to create a bit of a drag. They are then spun together in the opposite direction to that in which they were originally spun if unspinning the original work is to be avoided. On the wheel you do this by turning the wheel anti-clockwise instead of clockwise or vice-versa.

Dyeing

There is nothing more beautiful than natural, unbleached wool, but this or any other natural fibre can be dyed to delightful soft colours with natural dyes. It can be done before or after the fleece has been spun. The colour will vary, as the unspun fleece absorbs the colour unevenly, but this has advantages if you wish to card differently dyed fleece together to produce your own colours.

Mordanting

In order to dye with natural dyes the wool must first be mordanted. This is a process which allows the dye to be fixed to the wool. Mordanting (French 'mordre' – to bite) is a standard method whose details are given below. When the mordanting is complete the rest is creativity, and natural dyeing is certainly a field for experimentation.

There are five main mordants, and of these

generally three are used. The equipment needed for mordanting and dyeing is:

1   A large saucepan – aluminium, enamel or stainless steel are best.
2   Glass rods, sticks or wooden spoons for stirring.
3   Scales. These should measure down to 5 g. You can divide quantities up with a knife if you need smaller than you can measure.
4   A measuring cup or some glass jars *not* to be subsequently used for food.
5   A thermometer. This is especially needed for work with indigo and madder.
6   Stainless steel spoons for the chemicals.
7   Rubber gloves, an apron, muslin bags, rinsing buckets, scissors, a pen and notebook.

KEEP ALL DYEING EQUIPMENT SEPARATE FROM COOKING UTENSILS!

The procedure is as follows:

a)   Put the wool to be mordanted to soak in a plastic tub of warm water with a wetting agent. The wetting agent is a liquid which allows the fibre to be quickly penetrated to ensure even mordanting and dyeing.
b)   Two-thirds fill a large saucepan with water and warm up.
c)   Mix the mordant used with a half cup of warm water until dissolved, and add to the saucepan.
d)   Mix cream of tartar with a half cup of warm water until dissolved, and add to the pan, stirring well.
e)   Add the thoroughly wetted wool to the saucepan, submerge gently and slowly bring to the boil.
f)   Reduce to a simmer and continue for one hour.
g)   Remove from the pan, rinse in hot and then warm water, and hang up to dry.

The mordanting can be done at any time, preferably in large batches, as you then have the wool ready to dye whenever you want it.

Basic mordanting recipes

Quantities per 1 kg of wool.

(a)	Alum	225 g	60 g Cream of tartar
(b)	Chrome	15 g	15 g Cream of tartar
(c)	Tin	15 g	30 g Cream of tartar
(d)	Iron	30 g	60 g Cream of tartar
(e)	Copper	30 g	

Notes on mordanting

1 Keep each mordant clearly labelled.
2 Cream of tartar is used to minimize the damage done by mordants to the fibre and to encourage even penetration. It is not generally used when mordanting with copper.
3 The quantity of alum (aluminium potassium sulphate) shown in (a) above may be halved and, while the colours will not be quite so bright, they will be more intense. Exceptionally you *must* use full-strength alum with cotton.
4 Chrome (potassium dichromate) is light-sensitive when in solution, and therefore the pot must be covered during mordanting or uneven dyeing may result. It is also a good idea to dry the wool in the shade or dark and to store in a dark bag or cupboard. Too much chrome results in less colour fixing to the yarn.
5 When using tin (stannous chloride), which is incidentally very poisonous, it is important to reverse the order of adding to the dye pot. FIRST dissolve the cream of tartar and add it to the water in the pan and THEN add the dissolved tin. Adding the tin first will result in the formation of a white

gelatinous substance which will not mordant properly. The weight of tin used is critical.

6   Iron (ferrous sulphate) generally gives dark colours and, at a pinch, can be added at the end of the dyeing process to sadden a colour.

7   Copper gives a strong, green colour, but it is hard to overdye with it.

Additional chemicals which alter colours are acids such as vinegar (or diluted acetic acid) and lemon juice. These give redder colours generally but are rather unpredictable. Alkalis, which also change the colour, are bicarbonate of soda, washing soda and ammonia. You can also try salt, sugar and detergents. Mordanting does affect fibres, and certain mordants make the fleece more difficult to spin. The following is the order of increasing difficulty of spinning of mordanted fleeces:

chrome (easiest), alum, iron, tin, copper (most difficult).

Certain natural dyestuffs do not need to be mordanted before dyeing, as they usually contain tannin or aluminium salts within them. These are indigo (a vat dye needing additional chemicals to make up), walnut husks (the green outer shells), tea, coffee and lichens. The last mentioned is now a protected plant group, and their picking carries a large fine. They can also take a very long time to grow, but fortunately most of their colours can be found elsewhere.

To dye with indigo

Indigo makes an excellent blue, fast dye, but it has to be imported. The English equivalent is woad, which is very messy to work with and not nearly as efficient. No mordant is necessary, but a 'vat' solution has to be made which requires more precision and care than

97

most natural dyes. After dyeing the wool should be washed well and then it will be both wash and light fast.

To make vat indigo dye
60 g of indigo will dye 1 kg of wool.
Required:
   60 g indigo
   20 g caustic soda
   40 g dithionite (sodium bisulphate).

1 In order to make the dye soluble first mix the caustic soda to a paste with warm water (be careful, caustic soda is *very* corrosive and damaging to skin and eyes) then add to the indigo and mix to a slurry paste. The colour will change to a greenish-yellow. Mix well. Make this paste up to 1.25 litres of liquid; a scum will appear on top.

2 Sprinkle one third of the total dithionite into the indigo solution and mix in. The dithionite removes oxygen from the solution, which would otherwise change the nature of the dye from indigo blue to indigo white.

3 Gently heat the liquid to just below 40°C. The reaction generates heat and the temperature should be watched continually, additional heat being applied only when necessary.

4 Add the second third of the dithionite and stir it in. The solution will turn yellow and then change gradually to blue as the reaction between the oxygen and the dithionite proceeds. This will take about half an hour at the temperature suggested. In cold water the reaction takes place over three to four days and should take place in a long-necked, corked bottle which must be shaken twice a day.

5 When the reaction is complete half of the remaining dithionite should be added gradually while the

temperature is maintained at 40°C for one hour. A yellow coloration indicates that the reaction is over.

6   Cool the vat and pour into a long-necked bottle. Cork securely and label unambiguously.

7   The balance of the dithionite (i.e. one-sixth) is retained until dyeing occurs.

The ideal final solution has no blue particles floating in it and is a clear yellow colour. During storage a white sediment will form at the bottom of the bottle, and if this is warmed or shaken it will usually disperse. If not, a few drops of caustic soda solution will dissolve it. A bottled vat kept in a cool place will last at least two years.

To dye with vat indigo

1   Keep the temperature low throughout, between 50 and 60°C.

2   Fill a pot with water, and while it is heating thoroughly wet the cleaned yarn.

3   Sprinkle a few particles of dithionite on to the warming water and stir in to absorb any oxygen which may be dissolved.

4   Uncork the vat indigo bottle and remove a little dye, preferably by means of a small syringe inserted under the surface scum. Recork the dye bottle and add the dye to the warm water by putting the syringe nozzle under the water surface before emptying it. This will cut down the chance of air bubbles being introduced to the dye bath.

5   It is important to dye gradually for good penetration and fastness, as a lot of indigo can sit on the surface of the yarn. So be prepared to repeat the process several times.

6   Be sure that there is enough water in the bath for

the wool to move easily. About 2.5 litres of water are needed for each 50 g of wool.

7   When removing the wool from the dyeing bath try not to let the drainings drip back, because this reintroduces air to the dye. The wool must remain in the dye bath for only a few minutes. Drying between dips gives faster colours.

Clean rainwater usually is saturated with dissolved air and will therefore need the most dithionite. Uneven dyeing will be rectified after two or three dips, but if pale colours are required add very little indigo to the dye bath. Remember that light blue can be overdyed so easily. If the dyed wool is rinsed with vinegar it will make the dye faster.

Finally wash the yarn thoroughly with a mild, pure soap after dyeing to bring out the final colour.

Dyeing with natural dyes

Dyeing is very easy but, as with cooking, knowledge and success come with personal experience. It is vital to experiment and then keep explicit records of everything you did: time of day and year, type of water etc. It is difficult to repeat colours, but such records will help. There are very few precise instructions one can give.

When dyeing, tie up the wool first in hanks which have been secured loosely in three to four places – a figure of eight tie is suitable. Always dye on clean wool. The standard recipe for natural dyeing is equal weights of wool and plant, but with the imported dry dyestuffs such as logwood, madder and cochineal, less quantity is needed. Make sure the wool can move freely in the pot once all the plant material has been added. Put dyestuffs which are very fine (wood chips or small flowers) in a muslin bag to save trouble later in picking them all out. Dyes fade differentially in light and with washing, and the best way to test them is to hang samples in a sunny window and wash small amounts.

Try dyeing with any fruit, vegetable, flower, plant, root or bark growing around you and you will be astonished at some of the results. It is best to let the very tough materials such as roots or bark soak for at least twenty-four hours before dyeing with them, and indeed any plant is helped along with an overnight soak, although often it is not necessary.

Weaving

Having sorted the fleece, teased and carded the wool, spun the yarn, washed and fixed in the spin and completed the dye, you are now ready either to weave, knot or crochet the wool into the final product. This is the really exciting time, when you begin to see the fruits of your labours. Many specialized books are available, some of which are listed in Appendix I.

For weaving I suggest a simple frame loom, because it is versatile and uncomplicated to warp, and you can probably make it yourself. Even a good firm picture frame can be used, with nails spaced evenly top and bottom. Anything, in fact, which will be rigid and hold a steady tension.

For your first weavings use a linen or cotton warp, as your handspun will tend to stretch too much. Start with something smallish, or you will get bored with the design before you have finished it.

Knitting

When calculating the weight of unspun fleece you are going to need for a particular knitting project, bear in mind that there will be considerable loss in the processes of sorting, teasing, carding and washing. It can be up to half the starting weight depending on the quality of the wool and the amount of grease.

Normally knitting yarns are plied for strength, but it is not always necessary. Knit things in squares and oblongs using basic peasant designs. These look very good in handspun yarn which is in itself attractive and requires less tension adjustment than regular knitting patterns.

When you start making things you will find that your imagination will start to flood with ideas for things that you can make, and your hands will not work fast enough to fulfil all the ideas. Note them all down, because the work and the developing feeling for strengths and weaknesses of the materials stimulate the imagination, and these are the real ideas. How happy you will be when you finally sit on a rug that you have woven, dyed and spun from the wool of a sheep that you have shorn, cared for and maybe even helped into the world. And what a wealth of memories and insights into the secrets of existence it will hold.

A great Sufi (Arabic 'suf' – wool) saint was once asked: 'What is the secret of your intellect?' She replied: 'I spin at night by the candlelight, and when the candle fails I spin by the light of the moon.'

9 Health in the Flock

Sheep can be afflicted by a mass of ancient-sounding diseases and complaints, but there are now available numerous vaccines, inoculations, drenches, dips and pills which must have created an enormous change in shepherding. Of course these aids must never replace constant observation and attention to the sheep's needs. Neglected sheep are very vulnerable to ill health.

Healthy sheep
Head – the sheep should be bright-eyed and interested, with ears held firmly and not limply. Sheep hold their ears in differing positions according to their breed. The nostrils should be clean and very faintly dewy with no suggestion of stickiness, the mouth should be kept firmly shut except when cudding.
Respiration – the normal rate should be 20–30 per minute and can be counted while the animal is in repose, a respiratory movement being a complete rise and fall of the flanks and chest. In cold weather this is easily done by counting the condensed breaths. The rate will normally be higher when the animal is in pain, apprehensive or feverish, after exercise, or in

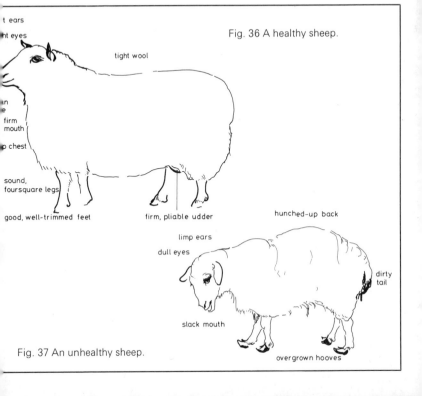

Fig. 36 A healthy sheep.

t ears
ht eyes
tight wool
an
e
firm mouth
p chest
sound, foursquare legs
good, well-trimmed feet
firm, pliable udder

Fig. 37 An unhealthy sheep.

hunched-up back
limp ears
dull eyes
dirty tail
slack mouth
overgrown hooves

very hot or cold weather. Young animals have a faster rate of respiration than adults.

Temperature – normally this is between 103–105°F.

Attitude – the healthy sheep stands with a straight back, foursquare on its legs, the tail hanging down straight and close to its body. When the animal is lying down it quickly responds to any noise.

The wool – this should never look wet or unreasonably straggly unless it is raining.

The dung – this is normally shiny, brown and pelleted; it falls away leaving a clean anus. The urine should be clear and pale yellow.

Cautionary signs

Head – hanging low with limp ears, or head jerking, teeth grinding, excessive salivation.

Respiration – extra fast and shallow.

Attitude – resting one leg, wriggling the tail, trembling, lack of interest, lying down with head outstretched, crows pecking near the body.

Wool – apparently wet, straggly, evidence of rubbing.

Dung – exceptionally loose and unformed. Urine red or cloudy.

Behaviour – sluggish movements, lagging behind the rest of the flock or even hiding itself away. This means that *it is essential* to count the flock at least twice a day.

Take every opportunity to study your charges. You will quickly realize what is normal behaviour. Learn to pick out the flock leader, to distinguish between their different bleats – yes, they are *all* different, and some much more noisy than others. Notice which sheep come first to the call and learn their pecking order. Any variation from the norm must immediately arouse your suspicions.

Grazing

The health of the flock can be maintained and even

improved by giving attention to the grazing available. Sheep living on the same area year after year are going to become riddled with worms. Grazing sheep tend also to be like horses, eating down the grass very low in some areas and leaving other regions untouched. This heavy grazing allows weeds such as buttercup, ragwort, daisies and plantain to flourish at the expense of the grazed and weakened grasses. Clover may spread, with the attendant risk of bloat. Controlled grazing, by dividing the available area into either permanent paddocks or temporary ones with hurdles or electric fencing, will allow the sheep to be moved as necessary. This will prevent the worm larvae from finding a host immediately and will also allow the grass time and peace to recover.

Mixed stocking, where different species either graze all together or after one another, is one more way of maintaining the health of the sward and controlling the worm population. Of course this is not always possible, but if it is, horses will graze after cattle, eating right up to their dunging areas, and then sheep can follow. Yet these in turn do not like the coarse grasses left by the horses but the cows will eat the grasses which have been left by both horses and sheep. The larvae and eggs left on the grass by the dung are destroyed when eaten by a different species.

Wounds
Sheep seem incredibly prone to getting caught up, torn and damaged if they become frightened and start running wildly. Normally they are very quiet and their thick fleece protects them – except of course when they have been newly shorn. During the winter any small cuts will heal easily, but in summertime the very real danger is from flies which lay eggs on the wound. These hatch within twenty-four hours into

maggots which burrow into the flesh causing great distress. Any wound must be covered with an antiseptic ointment which will also deter flies. This can be obtained from an agricultural merchant.

Wounds caused during sheep worrying by dogs must be treated with antiseptic antifly ointment and the sheep kept really quiet to allow it to recover from the shock and fright. Sheep tend to scatter wildly when chased by dogs, but are more likely to flock together when pursued by a sheepdog. Such an experience will often stop an animal eating and cudding, so see that it has fresh water and try to tempt it to eat. Ash leaves are often eaten by a wounded animal and they aid the healing process. Elm and elder leaves will also be favoured when all else is rejected. A mixture of leaves and grass can be cut up, pushed in the side of the mouth and encouraged down the throat with your fingers, and this may arouse the animal's interest in life once more. Avoid the sheep's back teeth as they are extremely sharp!

Fly-strike

As a small child, holidays on my uncle's farm seemed to be spent predominantly shepherding and dipping! Morning and early evening the sheep were slowly driven through a gap in some hurdles while they were counted and each animal scrutinized. I soon learned to recognize signs of fly-strike. Sheep blowfly (green not bluebottle) seem to home unerringly on to any sheep with a wound or break in the skin, however small. Sheep which are not thriving or have any degree of diarrhoea, however caused, also provide an ideal laying place for the fly. The eggs are laid under the dung or in a wound and hatch into maggots within forty-eight hours. These burrow into the skin and feed off the flesh.

Signs of maggots on a sheep are often barely discernible, at least in the early stages. The wool may be only *very* slightly darker in colour, but it may also look moist or even wet. The animal will often wriggle its tail more than usual, stamp its feet, shake its head and also hold it lower than usual. Sheep which have strike will also often *not* be doing as the rest of the flock. When the others are cudding the struck sheep will be wandering aimlessly, lying down and getting up again. Strike can also be smelled by the perceptive nose – once smelt never forgotten, for it is sickly sweet and extra 'sheepy'.

Treatment
Where eggs are found, any dirty wool must be cut away and the area rubbed thoroughly with fly ointment. If the eggs have hatched *each and every* maggot must be sought out and removed. If the skin is broken, press all round with your fingers and an incredible number of maggots of all sizes may often appear out of the smallest holes. Once struck a sheep will often be infested repeatedly unless steps are taken to dip the flock.

At least two dips for fly, and preferably three, are best carried out, one in late May or early June and then again in early autumn. Well-fed lambs on rich grazing are most likely to be affected owing to the possibly stronger smell of the dung and its possible looseness. Longwoolled sheep such as Lincoln Longwools are particularly prone before they are shorn. Regular summer dipping should reduce fly strike to a minimum. A moist, muggy summer day always makes me extra careful, as even a rain-wet fleece can entice the flies to lay.

It is vital to be really observant, as a struck sheep left for twenty-four hours can be nearly eaten alive. The shock of the damage can and does kill.

Feet

'No foot, no horse' applies to the sheep equally well. It is taken for granted that you buy stock with healthy feet from a flock where good foot hygeine is practised! Despite this, feet should be looked at frequently. Trimming is necessary to remove the horny growth which curves under the foot. If a smelly discharge is noticed remove all loose horn and expose the infected areas. Paint on a 4% formalin solution obtainable from the chemist and repeat weekly until the discharge ceases. When paring use foot trimming shears or a *sharp* hoof knife.

Catch your sheep by standing behind it and either side of its back feet. Put an arm under its neck and then sharply raise your arm and lift the sheep quickly so that it is sitting on its rump. If the animal is comfortable it will sit there indefinitely. You can now look at each foot and treat it as necessary. Pare the hoof gradually, cutting from heel to toe until a white line appears just inside the outside horn. Over-zealous trimming can make the foot bleed badly, and wounds so caused can allow infection such as footrot to enter or maybe even allow flies to strike. Commercial sheep farmers have permanent footbaths through which the sheep are run regularly once a

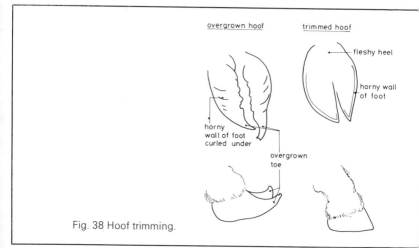

Fig. 38 Hoof trimming.

month. A home made footbath can be built. A long trough can be stood between four hurdles (two on either side), and the animals driven through. Fill with a mixture of 0.25 litres of commercial formalin and 4.5 litres of water.

Vaccination with Clovax coupled with routine trimming and footbaths will help control footrot.

External parasites

Ticks, keds and lice are all visible parasites which can cause irritation. They can be controlled by routine dipping which will also contain a fly-strike inhibitor, in a Ministry-approved sheep-dip. This should be done once in May and again in September-October time. Another parasite which is not seen by the human eye is the mite which results in sheep scab. The creatures burrow into the skin causing intense irritation, resulting in the animal rubbing off large areas of wool. It is obligatory for *all sheep* to be dipped (unless they are due for slaughter) in the autumn between two dates which are advertised nationally and in the local press. The dip used *must* be Ministry-approved to include control of scab, animals being kept in the dip for at least a minute and the head being pushed under at least once.

Dipping
Information can be obtained from your local animal health division of the Ministry of Agriculture, Fisheries and Food.

An old bath can possibly be used for small breeds, but for the larger ones a galvanized tank will suffice. Sometimes it is possible to dip your sheep with your neighbour's in his dip.

Points to be observed when dipping
Calculate the capacity of the dip to be used. Thor-

oughly clean the receptacle. Ensure either a soakaway which will not drain into any waterway or which will not allow animals or humans to gain access. We have our dip on the edge of a wood near the paddocks. We dug a 1 cubic metre hole, and filled it with gravel and broken up concrete into which the dip can be drained and which can be covered after drainage is complete. Alternatively you can arrange for used dip to be removed by a waste disposal company. *Always use a Ministry-approved dip.*

Mix the dip the day it is to be used, following the maker's instructions. If possible use rainwater to make up the dip, but failing this dissolve 1½ kg of washing soda to 450 litres (100 gallons) of water to soften it. If the dip needs to be replenished do so with the correct concentration of dip and water. DO NOT JUST ADD WATER. Try to choose a dry day. Dip in the morning to allow the sheep to dry before evening. Do not dip in wet weather. Do not dip sheep which have recently been fed. We keep our sheep in a stable overnight and give them access to drinking water only.

Stand the sheep to drain (we do this in hurdles beside the dip) in the shade for an hour, as they may scorch in the sun, before returning them quietly to the grazing area. Do not house the sheep while they are still wet, but do not allow them to be rained on excessively. These dips are *very* poisonous, so protective clothing must be worn. Rubber boots, rubber gloves and a facemask as well as an overall must be worn by all people directly concerned with the dipping. We invariably get drenched when dipping the sheep, so in spite of protection we like to shower after this operation!

Rinse all protective clothing thoroughly and dispose of dip canister through the local authority disposal service.

Fig. 39 Method of drenching.

Internal parasites

A rough guide to internal parasites present in your sheep can be gauged by collecting several dung samples and taking them to your vet for a worm count. There are several different sorts of worms which can affect sheep. The worm eggs are laid in the gut of the sheep and are passed out in the droppings. These hatch into larvae which, when eaten three to seven days later by the sheep while grazing, will reinfect them. Worms can be controlled by destroying them within the sheep, by drenching or injection, and also by constant change in grazing to cut down the chance of reinfection. Infection causes scouring, anaemia, general unthriftiness and possibly even death. Anti-worm medication can be administered by drench, which is given by mouth with the aid of a bottle or a drench gun. The cure may also be applied by injection. Both methods remove a wide range of important worms. When treating the sheep, pen to reduce activity and daubmark each animal with a coloured marking stick on its back as it is treated. Be sure to give the correct dose for age.

Drenching guns should be clean and checked for accuracy before starting. Point the nozzle over the tongue but not right down the throat. Let the fluid trickle in, holding the mouth shut to prevent the sheep regurgitating the liquid. Do not do this in the heat of the day as this may create an extra stress.

Injections can be given under the skin of the thigh or high up on the neck.

Treatment – dose or inject ewes in April or May (changing the grazing directly after the treatment). Dose the lambs in June as well, and every six weeks until September. Wet, warm summers provide ideal breeding conditions for worm larvae!

Scouring in lambs for no apparent reason may indicate a heavy worm burden, so treat the lambs and change the grazing as soon as possible after this is noticed. It is wise to treat any cattle, goats or pigs which share the grazing, once in spring and again in autumn. Doses will be indicated in the maker's instructions. Horses will need a completely different anti-worm preparation. Never vaccinate or inject a wet sheep as the risk of infection is much higher.

Clostridial diseases
There are a number of these diseases, which can all be treated with one multi-vaccine. 5 cc can be given to the yearling ewes before tupping; a month before lambing a further 2 cc is administered, followed by 2 cc just before lambing. This will protect the lambs for up to sixteen weeks. Older ewes will only need a 2 cc booster just before lambing.

The diseases which are covered by the vaccine are as follows. *Lamb dysentery*, in which unprotected young lambs fail to thrive, they may produce bloodstained dung, scour and soon die. *Enterotoxaemia* in adult sheep and *pulpy kidney* in lambs, which usually strikes the better animals in the flock which have been 'well done' (this term is used to describe an animal which has been well fed). First indication of this disease is the presence of a dead animal. *Braxy* or *Black disease* also produces sudden death (there is no treatment, but vaccination prevents it). *Tetanus*, which may enter any wound such as a newly docked tail or a shearing nick. It may kill, but mild cases will just exhibit a stiff walk and then the sheep recovers.

Mineral deficiencies
We hope that these will not be found in the small garden flock where the sheep have access to mineral licks and a mineral mix all the time. But in very high rainfall and sandy areas there may be a little more likelihood of deficiencies occurring. Discuss this with your vet. He should know the conditions current in your area and will advise you.

Diseases in the ewe
Pregnancy Toxaemia (twin lamb disease) – this occurs in undernourished ewes during late pregnancy when they are carrying two or more lambs. The ewe shows little inclination to move about and may appear blind or unsteady on her legs. She loses her appetite and may die within the week. The condition can be prevented by allowing the ewes constant access to good, leafy hay and fresh water. You may *think* they are not eating it much but sheep nibble constantly (but possibly not obviously) compared with cows and horses which eat in large mouthsful.

Hypocalcaemia – this is rather like milk fever in cows. Just before or after lambing the ewe will be seen to grunt, her legs will paddle and her ears will feel *cold*. If allowed to fall into a coma she will die. The vet can produce a quick response by injecting calcium borogluconate, but this must be done quickly after the first symptoms are seen. The condition may be aggravated by sudden changes in management and feeding just before lambing.

Mastitis – this is an inflammation of the udder. The acute form can be fatal. Avoid buying ewes with lumpy udders. The vet can treat a ewe with this disease in order to save the lamb, but the mother is best culled from the flock after weaning, as she may be a carrier of the infection.

Prolapse

This can happen both before or after lambing and is more likely to occur in well-done ewes carrying more than one lamb. The first indication of this particular disorder will be the appearance of the vagina, looking like a pink balloon, emerging from the vulva; in some cases the bladder also protrudes. Prolapse is most likely to occur any time up to three weeks before the expected lambing date. It is a condition needing professional help, but until the vet arrives keep the animal quiet in a cleanly littered pen. The parts will be replaced into the sheep and the vagina stitched, allowing space for the exit of urine, or a prolapse harness can be knotted on to the sheep. If the animal is sutured the stitch will have to be released as soon as the feet of the first lamb appear before the lambs are born. It is unwise to keep the ewe for another breeding season.

Prolapse of the uterus usually happens within hours of birth – the whole womb is produced, looking like a massive dark-red lump of flesh. The vet must be called as soon as possible to treat the animal. She may recover but cull her from the breeding flock.

Bloat

This can affect any sheep grazing on lush pastures, especially those containing clover. I find that late April and early May when the weather is thundery seems to produce more bloat in ruminants than any other period. The sulky sheep is also more liable to suffer. The animal bulges on its left side, looking rather like a balloon. Rapidly accumulated gas is not released by belching and paralyses the stomach nerves. The aim is to get the animal to belch to release this gas, and this can sometimes be done by tickling its throat. Failing this, always have a bloat drench ready. Prevention is better than cure, so before turning sheep on to lush pasture feed them some good hay, avoid rainy or dew-wet, rich pastures. Wait until they are dry and avoid *violent* changes in diet. Frothy bloat sometimes occurs where froth in the stomach blocks the gullet. We have experienced this on clovery grazing. A special frothy bloat drench will be needed.

Symptoms

Coughing – this is not normal in a healthy sheep, and worms must be suspected or perhaps dusty hay.

Diarrhoea – this may be caused by injudicious feeding of kale, cabbage, rapid changes in diet or more commonly, worms.

Constipation – this may be caused purely by lack of exercise.

Gid or sturdy

Tapeworm segments which have been passed in the dung of dogs or foxes amongst grass can be eaten by sheep. Inside the gut the eggs hatch into larvae and eventually enter the spinal column or the brain by way of the blood stream. There they enter the cystic stage, which may grow as large as a small apple. The cyst causes the sheep to hold its head on one side, walk jerkily or become paralysed. Your vet may be able to remove the cyst from the head. Make sure your dog is treated for tapeworms twice a year. Burn all its faeces for two days following the treatment.

Poisoning

Rhododendron – this causes the animal to vomit. Mix a teaspoon of bicarbonate of soda in a tablespoon of lard and push this on to the back of the tongue. This will make the animal vomit the leaves. Continue until

Table of poisonous plants and circumstances in which trouble is most likely

Horsetail	When present in hay
Bracken	When green or fresh cut
Buttercup	Only when alternative food is scarce
Kale	During pregnancy
St John's Wort	All stages cause photo-sensitization
Chickweed	Large quantities eaten by lambs
Oak	Fallen green acorns under the tree
Ragwort	All stages if eaten
Sheep's sorrel	Especially during lactation
Laburnum	Fallen seeds when ripe
Yew	Deadly poison at any time

no more pieces of rhododendron leaf are produced. A drench of strong tea or coffee should follow as a stimulant.

Kale – never feed more than $\frac{1}{2}$ kg per day, as it can in some cases cause acute anaemia. If so the urine will be red and jaundice may follow. Your vet may inject affected animals with vitamin B12 and iron. In spite of this treatment the animal may die.

Lead – this is found on old wooden, painted buildings. The animal will be found with its head turned back lying on the ground prior to convulsions. Four to six beaten eggs can be administered while professional aid is summoned.

Arsenic – this is present in some weedkillers and can poison. Get professional aid as quickly as possible.

Orf – this disease is a form of dermatitis which can infect humans as well as sheep. It is extremely painful in humans. It is caused by a virus, can appear at any time of year, and is more likely to affect lambs than adults. I have often seen it at sheep sales. Animals develop raspberry-like blisters which often break out on top of the foot and around the nose and mouth. This affects the suckling and grazing of the afflicted animals. Your vet will be able to provide treatment, but in areas where it seems common, vaccination will help to keep stock immune for up to twelve months. When handling infected sheep, wash hands carefully afterwards. I caught orf from a late orphan lamb one summer. The doctor seemed nonplussed by it, but eventually the local district nurse suggested Betnovate, which cured it. The pustules on my hand produced pain right up the arm!

Notifiable diseases and livestock movement record

The law makes it obligatory for owners of sheep and all other cloven-footed animals to keep a record of all movements on and off the premises. This is to enable the police and the animal health division of ADAS (the Agricultural Development and Advisory Service of the Ministry of Agriculture) to take control in the event of your vet diagnosing the following diseases:

Foot-and-mouth – this affects all cloven-footed animals. Possible symptoms are lameness, blisters on the claws, nostrils and possibly on the teats, excessive salivation and temperatures up to 41°C (106°F).

Anthrax – this is now a very rare disease which can also affect man. If you suddenly find a dead sheep inform the local divisional vet office and they will send a veterinary officer to investigate.

Scab – it is obligatory to dip with a Ministry-approved material between certain dates each autumn. The only sheep exempt are those being sold for slaughter within the period.

Rabies

From 1886–96, 2,807 cases of rabies were confirmed in Great Britain; of these seventy-eight occurred in sheep. With the recent awareness of the disease it is as well to know the symptoms.

The incubation period is between two weeks and three months. First there is a loss of appetite, followed by increased sexual desire. The sheep show little fear of water and bleat hoarsely. Eventually the animal is seen to strain constantly, becomes paralysed and dies after losing consciousness.

Over-eating

Keep all feed stores firmly shut. Pig rearing and fattening food can be fatal to sheep, as it contains too much copper for them. An extra feed of any grain or compound ration can cause death if ignored. If you *know* your sheep have raided the meal store get the sheep to the vet as soon as possible and he will treat them appropriately. We nearly lost a lamb this way. The trouble was that he continued to try the meal shed door daily till he was slaughtered.

Cast sheep

Sheep will end up on their backs unable to rise for many reasons, and if left in this state they will often die if left unaided for long. A sheep which *looks* as if she is resting may well be unable to rise and need help. Always closely check a sheep lying on its own.

Sheep with heavy or even wet fleeces may get onto their backs. A ewe heavy in lamb may lie down in a dip, or roll on her back, and be unable to get up again unaided. The meatier breeds and crosses can become cast far more easily than the less broad-backed mountain breeds or unimproved types. It has been suggested that sheep being deprived of mineral mixes may be more likely to become cast.

Veterinary aid

Professional aid is expensive, and where possible it will be cheaper to take the patient to the surgery. In some situations the vet will have to be called out, for instance in a serious case of prolapse. Some vets are only too pleased to discuss a problem and offer advice over the telephone. It is worth casting around your shepherding friends to find out which of your local vets is particularly interested in, and good with, sheep.

Bits and Pieces

Handling sheep

Sheep are *not* stupid, although poor nutrition may make them aggressive in their search for food. You only have to watch them to realize that they very quickly learn the time they are fed, moved or just looked at. They recognize the sight and sound of their handlers and have incredibly sharp hearing. Our sheep only have to hear the back door latch in winter and they will start to bleat for their next meal. A healthy sheep stands attentively watching when humans approach. If this is done slowly and quietly it will wait, sizing up the situation. If you move quickly the sheep will run – this is their natural means of self-preservation – but they soon stop to look around. If you learn always to move slowly they will trust you and respond by moving in the direction you either lead or drive them. You will find that they will not easily flock for humans but double back, dodge and then bolt! If necessary exploit your natural flock leader by getting her to lead the others; sheep tend to gravitate towards each other.

Sheep will quickly learn to follow a human. Always carry food which will rattle in your pockets and feed them as they approach. I know a shepherdess who always carries beans in her pocket. It is a magnificent sight to see three hundred or so ewes all galloping *towards* her as she approaches. She has very little difficulty in catching any of them.

You will notice that sucking lambs, weaned lambs and ewes with lambs all behave a little differently from empty sheep. When moving any sheep from one area to another it can be noticed that hungry sheep are more cooperative than full ones as they are probably expecting improved grazing. However any sheep can be taught to misbehave by bad management and foolish handling; they quickly discover the chinks in our armour.

A simple way to catch an individual animal is to throw some food down a metre or so away so that the one you wish to catch will eat with its *rump* towards you. Bend down and grasp a back leg above the hock with your opposite hand: it will struggle but soon give in. Slide the other hand under the neck and grasp the wool of the throat. If you are near a fence you can hold the animal against it with your knee against its shoulder. Another simple method for the taller,

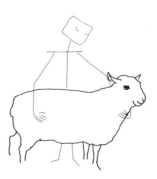

Fig. 40 Correct way to hold a sheep.

stronger person is to grasp the animal's loose skin in the groin as you lean over it, lifting the back leg off the ground.

Never handle a sheep by the wool on its body. A thin or unthrifty sheep will leave you with a handful of wool and a fat sheep will easily become bruised.

Casting sheep
You will need to learn this in order to check the feet. Try to get a shepherd to show you at first. Remember that heavy pregnant ewes are best not cast unless absolutely necessary. Handle them with extreme gentleness and care. Large breeds are heavy to handle and there are cradles available which will hold the sheep comfortably and safely for routine trimming. To cast and hold a sheep first catch it. With one hand under its neck and the other on the off side near its tail, you can gently push the head away from you, when it will roll over your knee and end up sitting comfortably between your feet. Avoid letting it sit on its tail but allow it to slump a little to one side or the other. Another method is then to allow the animal to lie on its side on the ground, putting your shin over its neck as you lean over it from the back. My uncle would often lay a sack over the sheep's head, and it would lie quite peacefully until it was removed. This is a handy way to trim a large bulky sheep's feet, as you will not have the heavy body nearly smothering you.

Docking lambs
The longer-woolled sheep will definitely need to be docked, although some backyarders seem to get away with not doing so. The removal of the tail cuts down the area of wool to be fouled by possible loose dung and consequent risk of fly-strike. The simplest method of docking is the Elastrator, used when the

lambs are two days old – the tails will usually drop off in a week or so. The apparatus can be obtained from agricultural merchants. The rubber ring is placed with the aid of the Elastrator pliers about 5 cm from the junction of the tail with the body between two vertebrae of the tail bone, leaving enough tail to cover the vulva of the ewe lamb. The ring cuts off the blood circulation and the tail beyond it dies. Mountain sheep are generally left with their tails on – there are usually no flies high up in the hills and the tail protects the udder.

Docking can be done with a very sharp knife. Holding the lamb between your knees cut the tail between two vertebrae leaving the correct length. The tail stump may bleed a little but a puff from an antiseptic powder will usually prevent infection and help stop the bleeding.

Castration

If you are eating your own males you may not bother to carry out this task, as they will often be fit for slaughter before the mating season. They will also be less likely to get overfat, as is possible with a wether lamb. The Elastrator can also be used to castrate ram lambs, but it must be done before they are two days old. The ring is placed over the scrotum, checking to see that both testes are present, and as close to the supernumerary teats as possible.

If the weather is at all cold keep the lambs in on clean straw, as they will probably be shocked and their resistance to disease lowered. Do not over-excite them. If the weather is really unsuitable you may have to postpone the operation and then use a Burdizzo (a bloodless) castrator. Get someone to give you a demonstration as *it is easy to operate incorrectly*. Lambs can be left until six weeks before castration by this method.

Shearing

The first time I sheared a sheep it took me two hours. I'm not much better now! This is a job for young men and women with strong backs!

Wool growth is affected by undue stresses during pregnancy, poor nutrition at any time, heavy worm infestation and consequent possible disease. The good shepherd aims to reduce these stresses to a minimum in order to get a good fleece. Eighty per cent of the growth occurs between July and November. The shearing may be done in very late May if the weather is warm, but more usually in early June when chills are unlikely, and preferably before the really hot weather sets in. The lambs will also be old enough not to be too disturbed by the handling of their mothers.

The fleeces are ready if you can see the yolk, a yellow fluid produced by the sebaceous glands just under the skin. This material is the lanolin which gives the wool its greasy feel, and it should have risen at least 8 mm ($\frac{1}{4}$ in) up the wool and away from the body. This fluid seems to rise when the warm weather comes. The shearing is difficult before it has risen enough for you to get the shears between the yolk and the flesh of the sheep. The fleece will also need to be clean and free from twigs and vegetation. It is wise to 'dag' or remove any soiled wool from around the tail area a few days prior to shearing.

It is difficult to shear a wet sheep, apart from the fact that the fleece will heat after it has been sheared and so lose quality. Choose a day when the weather is warm but not too hot, and wait till the dew is off the sheep's backs.

Pen the animals fairly closely so they can't move about too much and are also in the shade in a clean area near to your shearing spot. The ideal place is a covered area with a clean, wooden floor. You will need soft footwear – the professionals make a quick moccasin out of sacking tied with binder twine; these will enable you to move quickly and quietly about the shearing floor. A boiler suit or dungarees are ideal, as shearing is a greasy, smelly job!

Tools – if you only have two or three sheep the work can be done with dagging shears, but it is a long job. Old hand-operated clipping machines, often go for a song at sales. If renovated and sharpened they can be good, but you will need someone to turn the handle. Electric shears are the tools of the modern shearer, and perhaps you can borrow these.

It is wise to watch someone shearing before embarking on your own task. The professionals aim to clip at least 150 sheep a day!

Your shearing must be positive, as weak movements produce a pecking action which results in second cuts which spoil the fleece, and time is lost by cutting twice. Keep the shears close to the flesh but avoid nicking the skin by pulling the skin tight. Never hold the skin ahead of the shears away from the sheep as you will nick her. Keep the points of the shears in the wool at the end of a strike or 'blow'.

Catch your sheep putting your arm under the throat while your feet are on either side of its back feet, then as you walk back sharply lift your arm and the sheep will sit on the side of its rump. You must be prepared to shear with either hand or your shearing hand will become very tired. The free hand will be behind the shears and on the bare skin. Clip close to the skin to avoid bad ridges, but hand clipping does leave more marks than machine clipping. Start around the right ear and the cheek and then move down the neck and brisket. This 'opens' the wool. You will then cut up the back leg to the backbone and continue these blows from belly to backbone, trying to keep the fleece in one piece. The other side is then started from the other back leg up to the backbone

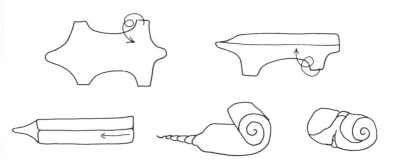

Fig. 41 Method of rolling a fleece.

with long blows avoiding second cuts until the whole fleece falls off. Allow the shorn sheep to run into a separate pen where they can be given feed and water. They will need protection against rain for at least a week after shearing.

Rolling the fleece – make sure your floor is free from debris. Lay the fleece flesh side down, except for mountain breeds, which are rolled flesh side inside. Pick off any vegetation and then fold the flanks in and start rolling from the tail end up to the neck. As you get to the neck pull this wool into a string, twisting gently to make a loose rope to tie up the fleece. Avoid using string or binder twine as this gets tangled in the wool.

Wool Marketing Board – all wool producers keeping more than four sheep must register with the Wool Marketing Board. You can arrange with the WMB to collect the wool or you may want to spin this yourself.

Storage – keep the fleeces safe, as they are valuable. Do not cover with plastic or they may sweat, but a plastic sheet under them will protect from damp and dirt. If selling, the WMB will supply wool sacks in which to pack the fleeces, but remember to label them with your name. Daggings and loose wool should be stored separately. You will be paid on quality and weight of fleece.

Milking sheep
Milking a sheep is much like milking any other animal, except that the teats are smaller than most and there are only two of them. It will be a case of milking with one finger and thumb, and this must be done very gently and rhythmically. The milk contains a relatively high butter fat content, usually 6–7% compared with 3–5% in the average cow. It makes good yoghurt and also cheese.

A simple cheese can be made by hanging yoghurt up to drain in a muslin bag for twenty-four hours and then scraping down the curd every two or three hours. The whey is excellent for cooking and may be used instead of milk for scones and bread. Mix the strained curd with pepper and salt to taste, store in a fridge and eat within a week.

You may have to milk a bereaved sheep if you have no spare lambs. She will quickly acquiesce if you are firm at first. Make sure that she is held immobile by someone else while you milk. You can milk her twice daily like any milk animal, but she will not give as much milk as the milk breeds kept on the Continent and in the Middle East.

Identification
This can be done by means of *ear tags*, which may be numbered, lettered and are variously coloured. Daltons supply tags coloured and marked as requested with the appropriate applicator. The tag generally goes into the sheep's left ear.

The sheep's ear can be *tattooed* on the inside with the aid of a tattoo punch. This is often used for pedigree animals, but is of course not so visible as tags.

Washable dye can be painted on the sheep's wool. Large numbers are easily identifiable painted on either side of the sheep to coincide with ear tattoo marks.

General hints
Insurance
You may feel it is wisest to insure your stock against trespass or causing accidents if they get out on to the road. Fencing is never foolproof, however good.

Fig. 42 Many common hazards can be easily avoided.

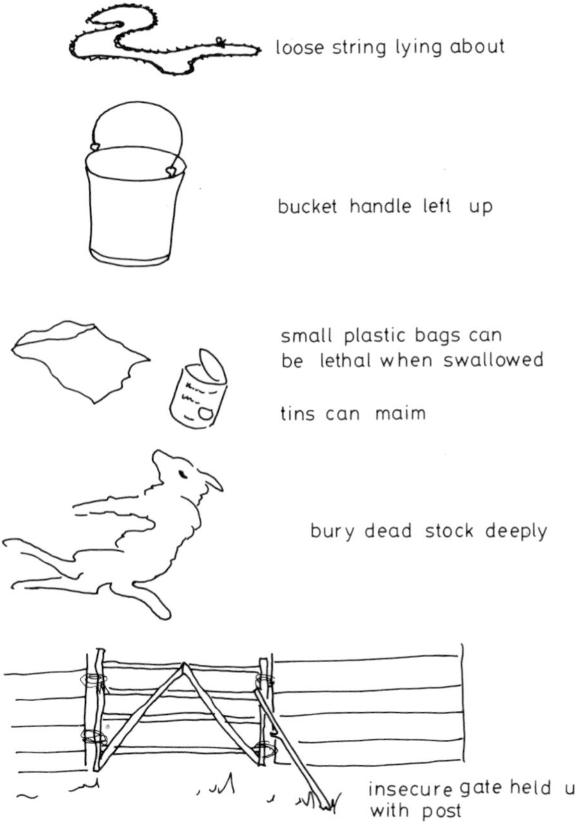

loose string lying about

bucket handle left up

small plastic bags can be lethal when swallowed

tins can maim

bury dead stock deeply

insecure gate held up with post

Pets and sheep

Sheep do not appreciate grazing after chickens or geese and will spurn the grass which has been walked on continuously by dogs, cats and humans. We must always realize that grass is their complete diet for many of the warmer months of the year, so try and keep the children and pets off the sheep areas.

I find sheep are a continual pleasure (although they might be considered smelly by some), time-consuming, charming, aggravating, amusing and even affectionate. They will certainly need patience and diligence; they will also improve your perception of both stock and the weather. You will find you are seriously studying the sky and the cloud formations which may indicate changes in the weather, and will notice the signs of forthcoming frost, wind or rain and so take appropriate action. Of all stock I think sheep seem to be the most reliant on our greatest possible attention and observation. Time spent just watching your sheep is never wasted. As you become more sheep minded you will be able to nip trouble in the bud before it develops. It is always worth taking some sort of action even if you only *feel* there is something wrong. Catch the particular sheep to check it, or just *watch* it to see if it really *is* still eating and not just standing mutely with its head lowered while the others graze.

Appendix 1

Further Reading

Periodicals

The Sheep Farmer, published quarterly by National Sheep Association

Practical Self-Sufficiency, published bi-monthly by Broad Leys, Saffron Walden, Essex

The Ark, published monthly by Countrywide Livestock Ltd, Winkleigh, Devon, EX19 8SQ

Crafts Magazine, published bi-monthly by Crafts Advisory Committee, 12 Waterloo Place, London SW1Y 4AU

The Weaver's Journal, published quarterly by Journal of Weavers, Spinners and Dyers, c/o Federation of British Craft Societies, 80a Southampton Row, London WC1B 4BD

Books:

British Sheep – illustrated guide to breeds, National Sheep Association

British Sheep Breeds, Their Wool and Its Uses, The British Wool Marketing Board

Sheep Production, Allan Frazer, Nelson

Sheep Farming Today, J. F. H. Thomas, Faber

Sheep Production and Grazing Management, C. R. W. Spedding, Baillière, Tindall & Cassell

The Shepherd's Guide Book, Margaret Bradbury, Rodale Press

The Backyard Sheep Book, Ann Williams, Prism Press

T.V. Vet Sheep Book, T.V. Vet, Farming Press

Herbal Handbook for Farm and Stable, Juliette de Baïracli Levy, Faber and Faber

British Poisonous Plants, HMSO Bulletin No. 161

Electric Fencing, HMSO Bulletin No. 147

Wool Away – The Art and Technique of Shearing, Godfrey Bowen, Whitcombe and Tomb Ltd, New Zealand

Farming Organically
Self-sufficient Smallholding ⎫ all available from the
Smallholder's Harvest ⎬ Soil Association
Looking at Livestock ⎭

Your Handspinning, Elsie G. Davenport, Select Books

Your Yarn Dyeing, Elsie G. Davenport, Select Books

The Magic of Spinning, Marion Channing, Marion Channing

Spinning and Weaving in Palestine, Shelagh Weir, British Museum

Notes from the Bankfield Museum, H. Ling Roth, Bankfield Museum

Handweaving and Cloth Design, Marianne Straub, Pelham Books

Handspinning, E. Leadbeater, Studio Vista

Creative Crafts with Wool and Flax, Molly Duncan, Bell

Spin Your Own Wool, Molly Duncan, Bell

Cut My Cote, Royal Ontario Museum

The Dyer's Companion, Elijah Bemiss, Dover

The Use of Vegetable Dyes, Violetta Thurstan, Dryad

Natural Dyes and Home Dyeing, Rita Adrosko, Dover

Dyes from Natural Plants, Anne Dyer, Bell

Weaving Is Fun, A. V. White, Dover

The Technique of Woven Tapestry, Tadek Beutlich, Batsford

Weaving, N. Znamierowski, Pan

The Traditional Moroccan Loom, C. F. McCreary, Thresh, USA

Sheep and Wool for Handicraft Workers, Michael L. Ryder, available from The Ark, Winkleigh, Devon EX19 8SQ

Appendix 2

Useful Addresses

National Sheep Association, Jenkins Lane, Tring, Herts
British Wool Marketing Board, Kew Bridge Road, Brentford, Middlesex TW8 0EL
Rare Breed Survival Trust Ltd, P. F. Hunt, 127 Abbots Road, Abbots Langley, Herts WD5 0BJ
The Soil Association, Walnut Tree Manor, Haughley, Stowmarket, Suffolk IP14 3RS

Suppliers of equipment:

Central Wool Growers, Wool & Agricultural Merchants. Branches in Stamford, Towcester, Market Rasen, Kentford, Dereham (Norfolk)
Dalton's Supplies Ltd, for ear tags, Nettlebed, Henley-on-Thames RG9 5AB
Gallagher Electronics Ltd (free illustrated guide to electric fencing) PO Box 5324, Hamilton, New Zealand
   UK distributor, Rutland Electronics (Oakham), 31 High Street, Oakham
Hunters of Chester, Grass & Clover Seed Specialists, Chester
H. S. Jackson & Son (Fencing) Ltd (free illustrated fencing guide) Stowting Common, Near Ashford, Kent TN25 6BN
Self Sufficiency Supplies Ltd, Priory Road, Wells, Somerset
Small Scale Supplies, Widdington, Saffron Walden CB11 3SP

Spinners' and weavers' supplies:

British Wool Marketing Board, Oak Mills, Station Road, Clayton, Bradford, W. Yorks BD14 6JD — Fleeces

County Craft, 22 Market Street, Alton, Hants GU34 1LB — Spinning and weaving equipment, yarns and fleeces, books

Craftsman's Mark Ltd, Trefaant, Denbigh LL16 5UD, North Wales	Natural yarns, textile study notes
Campden Weavers, 16 Lower High Street, Chipping Campden, Gloucestershire	Yarns, books, spinning and weaving equipment
Dryad, PO Box 38, Northgates, Leicester, LE1 9BU	School suppliers of looms etc.
Frank Herring and Sons, 27 High West Street, Dorchester, Dorset DT1 1UP	Weaving and spinning equipment
K. R. Drummond, 30 Hart Grove, Ealing Common, London	Books only
Guernsey Weavecraft, Juniper Cottage, Belmont Road, St Peter Port, Guernsey	Yarns, looms and accessories
The Handweavers Studio and Gallery, 29 Haroldstone Road, London E17	Weaving and spinning equipment classes
Kennet Wool Workshop, 77 High Street, Marlborough, Wiltshire, SN8 1HF	Spinning and weaving equipment classes
Eliza Leadbeater, Rookery Cottage, Dalefords Lane, Whitegate, Northwich, Cheshire	Weaving and spinning equipment
Muswell Hill Weavers, 65 Rosebery Road, London N10 2LE	Natural dyes, courses in spinning and dyeing
Olicana Crafts, Unit 27, Piece Hall, Halifax	Wheels and other equipment
The Strangers Wool Shop, 5 Timber Hill, Norwich, Norfolk	Spinning and weaving equipment classes
Weavers Dream, Barton-St-David, Somerton, Somerset	Fleeces, spindles, some looms

Appendix 3

Fifty Plants for Natural Dyeing

Plant Material	Part Used	Source	Mordant	Colour	Fastness
Agrimony	leaves, stalks	fresh	alum, chrome	yellow	good
Alkanet	roots	dried	none	light blue	good
		imported	alum	reddish brown	good
Beetroot	leaves	fresh	alum	light yellow	poor
	roots	fresh	alum	light orange	poor
	roots	fermented	alum	stronger orange	fair
Blackberries	berries	fresh	iron	purple grey	fair
Broom	flowers	fresh, dry	alum, chrome	yellow	good
Carrot	leaves	fresh	alum, chrome	yellow, tan	good
	roots	fresh	alum, chrome	pale greens	good
Chamomile	flowers, small	fresh, dry	iron	fawn	good
	leaves		alum	yellow	good
Cloves	whole	dry	alum	cream	fair
Cochineal	beetles	dried, imported	alum	strong pink	very good
				light red	
			tin	bright red	very good
			chrome	purples	very good

Plant Material	Part Used	Source	Mordant	Colour	Fastness
Cochineal *cont.*			copper	sandy purple	very good
			iron	pinky grey	very good
Coffee	grounds	fresh or used	tin, chrome	gold	good
Cutch	wood	powdered, imported	alum, tin	browns	good
Dahlia	flower heads	fresh, dry	no mordant	tans	good
			alum, chrome	yellow, but varies with flower colour	good
Dandelion	heads	fresh, dry	alum	light yellow	good
	roots	fresh	chrome	pale green	fair
				yellow, said to produce magenta	
Dock	leaves	fresh, dry	chrome	green brown	good
	roots	fresh	alum	dull yellow	fair
Elderberries	berries	fresh	alum	light purple	fair
	leaves	fresh	iron	blue grey	fair
			chrome	light brown	fair
			vinegar, salt	more lilac	fair
Forsythia	flowers	fresh, dry	alum, tin	bright yellow	good
Fustic	bark	dried, buy	alum, chrome	yellows	good
Golden Rod	flowers	fresh, dry	alum	bright yellow	good
			chrome	bronze	good
Golden Seal	root	dried	tin	gold/yellow	good
Gorse	flowers	fresh, dry	alum	light yellow	good

Plant Material	Part Used	Source	Mordant	Colour	Fastness
Heather	flower tips, leaves	fresh, dry	alum	bright yellow	good
			copper	green gold	good
			iron	green	good
Henna	leaves	dried, powdered	tin, alum	reddish brown	good
Hops	leaves, flowers	dried	alum, chrome	beiges	fair
Horse Chestnut	leaves	fresh	alum	gold brown	good
			iron	warm grey	good
				fawn, yellows	
Horsetail	whole plant except root	fresh	alum	yellow	good
Indigo	leaves	buy powdered	not soluble in water – needs special preparation	blue, all shades	very good
Ivy Berries	berries	fresh, dry	chrome	brown green	good
		ripe	tin	grey pink	good
		unripe	alum	pale yellow	good
Juniper Berries	berries	dried	copper	khaki	good
Lady's Bedstraw	tops	fresh	alum, chrome	yellow	good
	roots	fresh	alum, chrome	reds	good
Logwood	barks, chips	dried	chrome	black	very good
		imported	alum, iron	purples	very good
			tin	bright purple	very good
Madder	root, powdered	dried	alum	light orange	very good

135

Plant Material	Part Used	Source	Mordant	Colour	Fastness
Madder cont.	chopped	imported	chrome	coral	very good
			tin	rust, sandy	very good
				dark salmon	
Marguerite	flowers	fresh, dried	alum, chrome	yellows, gold	good
Marigold	flowers	fresh, dried	tin	light yellow	good
Meadow Sweet	whole plant	fresh	alum	greenish yellow	fair
	roots	fresh	alum	red	good
			chrome	chocolate brown	fair
Nettle	flowers	fresh	alum	dull gold	good
	whole plant	fresh	alum	greenish yellow	good
Oak	bark	collect dry	no mordant	black	good
Onion skins	brown outer	dried	alum	gold	good
	leaves		chrome	rust	very good
Pigweed	whole plant	fresh, dry	alum, copper	moss green	fair
Pine	cones	dried, fallen	alum	dull brown	fair
			no mordant	light yellow	fair
Pomegranate	skins	dry	alum	yellow green	good
			iron	brown	good
Privet	branch tips	fresh	alum, chrome	yellow gold	good
Red Cabbage	outer leaves	fresh	chrome	light green	poor
St John's Wort	flowers, tops	fresh, dried	copper	light gold, browns	good
			alum	gold	good
			chrome	gold	good

Plant Material	Part Used	Source	Mordant	Colour	Fastness
St John's Wort cont.	flower buds		vinegar	said to give crimson	good
Sanderswood	chips	dried	chrome	light grey, pink	fair
			alum	oranges, salmon, sandy	good
Sassafras	bark	dried	chrome	rose brown to burnt orange	good
Tea	leaves	dry	copper	pale brown	good
			iron	grey	good
Turmeric	root, powder	dried	alum	yellow	fair
			chrome	gold	fair
Walnut	shells	dried	tin, no mordant	cream	fair
	husks	fresh	copper	good brown	good
Weld	whole plant	fresh, dry	alum	yellow	very good
Woad	leaves	fresh picked requires a special fermentation process	no mordant	blues	very good

Index